D0026567

Compressor Handbook: Principles and Practice

By
Tony Giampaolo, MSME, PE

THE FAIRMONT PRESS, INC.

CRC Press
Taylor & Francis Group

Library of Congress Cataloging-in-Publication Data

Giampaolo, Tony, 1939-
　　Compressor handbook: principles and practice/by Tony
Giampaolo.
　　　　p. cm.
　　Includes index.
　　ISBN-10: 0-88173-615-5 (alk. paper)
　　ISBN-10: 0-88173-616-3 (electronic)
　　ISBN-13: 978-1-4398-1571-7 (Taylor & Francis : alk. paper)
　　1. Compressors--Handbooks, manuals, etc. I. Title.

TJ990.G53 2010
621.5'1--dc22   2010008818

Compressor handbook: principles and practice by Tony Giampaolo
©2010 by The Fairmont Press. All rights reserved. No part of this publication may be reproduced or transmitted in any form or by any means, electronic or mechanical, including photocopy, recording, or any information storage and retrieval system, without permission in writing from the publisher.

Published by The Fairmont Press, Inc.
700 Indian Trail
Lilburn, GA 30047
tel: 770-925-9388; fax: 770-381-9865
http://www.fairmontpress.com

Distributed by Taylor & Francis Ltd.
6000 Broken Sound Parkway NW, Suite 300
Boca Raton, FL 33487, USA
E-mail: orders@crcpress.com

Distributed by Taylor & Francis Ltd.
23-25 Blades Court
Deodar Road
London SW15 2NU, UK
E-mail: uk.tandf@thomsonpublishingservices.co.uk

Printed in the United States of America
10 9 8 7 6 5 4 3 2 1

0-88173-615-5 (The Fairmont Press, Inc.)
978-1-4398-1571-7 (Taylor & Francis Ltd.)

While every effort is made to provide dependable information, the publisher, authors, and editors cannot be held responsible for any errors or omissions.

Pages 221-232: Compressor specifications in Appendix XX From the 2009 Compressor Technology Sourcing Supplement, courtesy COMPRESSORTechTwo magazine, published by Diesel & Gas Turbine Publications. Current compressor information can be found at www.compressortech2 or www.CTSSNet.net.

Dedication

This book is dedicated to my five grandchildren:

Amanda Rose
Anna Josephine
Carly Paige
Riley James
Nickolas Anthony

They are the future.

Contents

APPENDIX

Preface

Compressors have played a major role in setting our standard of living and they have contributed significantly to the industrial revolution. Early compressors like the bellows (used to stoke a fire or the water organ use to make music) marked the beginning of a series of compression tools. Without compression techniques we could not have efficiently stabilized crude oil (by removing its trapped gasses) or separated the various components of gas mixtures or transported large quantities of gas cross country via gas pipelines. Today, compressors are so much a part of our every day existence that many of us do not even recognize them for what they are. Compressors exist in almost every business and household as vacuum cleaners and heating & air conditioning blowers. Even those who have worked with compressors (usually only one or two types of compressor) have only a vague awareness of the variety of compressors in existence today.

It is always interesting to see how the inventive process takes place, and how the development process progresses from inception to final design. Therefore, included in some sections of the book is historical information on the development of various compressors. Due to the number of different types of compressors it was too time consuming to research the origins of each compressor type. For the roots blower and screw compressor the inventive process is clear as discussed in Chapters 1 and 3. However, the origin of the reciprocating compressor is somewhat obscured. No doubt the water organ devised by Ctesibius of Alexandria paved the way. Nevertheless, using water to compress air in a water organ is a far cry from a piston moving within a cylinder to compress gas. True there is significant similarity between reciprocating engines and reciprocating compressors: Just as there is similarity between turbo compressors and turbine expanders.

Many engineers / technicians / operators spend their entire careers in one product discipline (manufacturing, maintenance, test, sales, etc…). Sometimes they have had the opportunity to work in several disciplines. This book is intended to assist in the transition from an academic background to a practical field, or from one field to another. It will assist the reader in his day-to-day duties as well as knowing where to look for additional information. Also people respond better when they understand why they are asked to perform certain functions, or to perform them in a certain order.

My intention is to provide a basic understanding of the variety of compressors. The need for this book has grown out of the request for seminars and training sessions from utilities and oil & gas companies. Most often these companies hire new employees or relocate and retrain their current employees. The reader may have some experience in the operation or maintenance of some compression equipment from previous assignments.

This book provides a practical introduction to dynamic and positive displacement compressors, including compressor performance, operation and problem awareness. In reading this book the reader will learn what is needed to select, operate and troubleshoot compressors and to communicate with peers, sales personnel and manufacturers in the field of dynamic and positive displacement compressor applications.

In addition to the theoretical information, real life case histories are presented. The book demonstrates investigative techniques to identify and isolate various contributing causes such as: design deficiencies, manufacturing defects, adverse environmental conditions, operating errors, and intentional or unintentional changes of the machinery process that precede the failure. Acquiring and perfecting these skills will enable readers to go back to their workplace and perform their job functions more effectively.

In addition to the content of this book the engineer/technician/ operator will find that the information provided in the appendix will become a useful reference for years to come.

Tony Giampaolo
Wellington, FL
January 2010

Acknowledgements

I would like to recognize and thank the following individuals for their support and assistance in obtaining photographs for use in this book:

Norm Shade, President, ACI Services, Inc.
Danny L. Garcia, Project Manager, Sun Engineering Services, Inc.
Roger Vaglia, Product Manager (Retired), Cooper Ind., White Superior Division
John Lunn Engineering Manager (Retired) Rolls Royce USA
Everette Johnson, Engineering Manager, Cameron Compressor Corporation
Ben Suurenbroek (Retired), Cooper Energy Services—Europe
Dave Kasper, District Manager, Dresser Roots, Inc.

I also want to acknowledge and thank Peter Woinich, Design Engineer, Construction Supervisor and Associate (Retired) of William Ginsberger, Associates for his help in proofreading this manuscript.

Also I wish to acknowledge and thank the following companies for their confidence and support by providing many of the photographs and charts that are in this book.

ACI Services, Inc.
Baldor Electric Company
Cameron Compressor Corporation
COMPRESSORTech^Two magazine, published by Diesel & Gas Turbine Publications
Dresser Roots, Inc
Gas Processors Suppliers Association
MAN Turbo AG
Oil & Gas Journal
Penn Engineering
Petroleum Learning Programs
Rolls Royce USA
Sun Engineering Services, Inc.
United Technologies Corp, Pratt & Whitney Canada

Chapter 1

Introduction

HISTORY

The history of compressors is as varied as are the different types of compressors. Therefore it is fitting that we first identify the different types of compressors. As shown in Chart 1-1, compressors fall into two separate and distinct categories: dynamic and positive displacement.

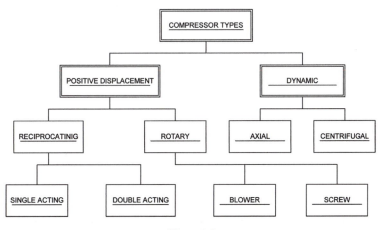

Chart 1-1

Somewhere in antiquity the bellows was developed to increase flow into a furnace in order to stoke or increase furnace heat. This was necessary to smelt ores of copper, tin, lead and iron. This led the way to numerous other inventions of tools and weapons.

One of the earliest recorded uses of compressed gas (air) dates back to 3rd century B.C. This early use of compressed air was the "water organ." The invention of the "water organ" is commonly credited to Ctesibius of Alexandria[1]. The concept was further improved by Hero of Alexandria (also noted for describing the principles of expanding steam to convert steam power to shaft power).

1

The water organ consisted of a water pump, a chamber partly filled with air and water, a row of pipes on top (organ pipes) of various diameters and lengths plus connecting tubing and valves. By pumping water into the water/air chamber the air becomes compressed. Than by opening valves to specific organ pipes the desired musical sound is created.

Ctesibius also developed the positive displacement cylinder and piston to move water.

It was not until the late 19th century that many of these ideas were turned into working hardware.

In the 1850s, while trying to find a replacement for the water wheel at their family's woolen mill, Philander and Francis Roots devised what has come to be known as the Roots blower[3]. Their design consisted of a pair of figure-eight impellers rotating in opposite directions. While some Europeans were simultaneously experimenting with this design, the Roots brothers perfected the design and put it into large-scale production.

It is not surprising that other compressor designs followed power-

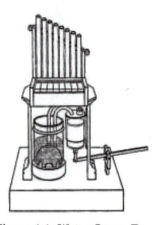

Figure 1-1. Water Organ Developed By Ctesibius.[2]

Figure 1-2. Photo Courtesy of Frick by Johnson Controls.

Figure 1-3. Cooper-Bessemer Z-330 Integral Engine Compressors in Krunmhorn, Germany. Courtesy of Ben Suurenbroek (Retired Cooper Energy Services)

producing designs. For example, the reciprocating engine concept easily transfers to the reciprocating compressor.

The integral-engine-compressor is a good example as its design utilizes one main shaft connected to both the power cylinders and the compression cylinders. The form and function of the compressor cylinders are the same whether it is configured as an integral engine-compressor or a separable-compressor driven by an electric motor, gas engine or turbine.

Other examples are the centrifugal compressor, (Figure 1-4) the turbo-expander, the axial compressor, and the axial turbine (Figure 1-5 and 1-6).

In 1808 John Dumball envisioned a multi-stage axial compressor. Unfortunately his idea consisted only of moving blades without stationary airfoils to turn the flow into each succeeding stage.[4,5,6]

Not until 1872 did Dr. Franz Stolze combine the ideas of John Bar-

Figure 1-4. Barrel Compressor Courtesy of Rolls-Royce USA (formerly Cooper Industries Energy Services).

Figure 1-5. Five Stage Power Turbine Rotor From RT15 Turbine Designed For 12,000 RPM Courtesy of Rolls-Royce USA (formerly Cooper Industries Energy Services).

ber and John Dumball to develop the first axial compressor driven by an axial turbine. Due to a lack of funds, he did not build his machine until 1900. Dr. Stolze's design consisted of a multi-stage axial flow compressor, a single combustion chamber, a multi-stage axial turbine, and a regenerator utilizing exhaust gases to heat the compressor discharge gas. This unit was tested between 1900 and 1904, but never ran successfully.

Operating conditions have a significant impact on compressor

INTRODUCING THE ST18

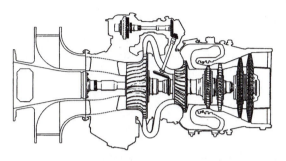

2MW AERODERIVATIVE

Figure 1-6. Courtesy of United Technologies Corporation, Pratt & Whitney, Canada. The ST-18 is a 2 Megawatt Aeroderivative Combining Centrifugal Compressor & Axial Expansion Turbine.

selection and compressor performance. The influences of pressure, temperature, molecular weight, specific heat ratio, compression ratio, speed, vane position, volume bottles, loaders and unloaders, etc. are addressed in this book. These conditions impact compressor capacity and therefore the compressor selection. They also impact the compressor efficiency. Flexibility in selection is still possible to some extent as compressors can be operated in parallel and series modes. For example, to achieve higher pressures multiple compressors can be configured in series whereby the discharge of one compressor feeds directly into the suction of a second compressor, etc. Likewise, to achieve higher flows multiple compressors can be configured in parallel whereby the suction of each compressor is manifolded together and the discharge of each compressor is also manifolded together.

Different methods of throughput control are addressed in Chapter 5, such as, discharge throttling, suction throttling, guide vane positioning, volume bottles, suction valve unloaders and speed control; and how each of these control methods effects compressor life.

This book discusses different compressors; how they operate and how they are controlled. Since the cost of process downtime and damage to a compressor can range from thousands to millions of dollars; the types of failures that can occur and how to avoid these failures is also addressed in this book.

In view of the fact that the most destructive event in a dynamic compressor is surge, compressor surge will be defined and discussed in detail. Also discussed are the various types of instrumentation (controllers, valves, pressure and temperature transmitters, etc..) available and which are most suitable in controlling surge. Destructive modes of other compressors are also addressed.

A few algorithms are presented, primarily in Chapters 4 and 7, to help demonstrate interactions of pressure, temperature and quantify results, but their understanding is not essential to the selection of the proper control scheme and instrumentation. The reader should not be intimidated by these algorithms as their understanding will open up a broader appreciation of how the compressor works.

Footnotes
1   *A History of Mechanical Inventions*, Abbott Payton Usher. This Dover edition, first published in 1988, is an unabridged and unaltered republication of the revised edition (1954) of the work first published by Harvard University Press, Cambridge, MA, in 1929.
2   Multiple sources were found for this sketch, none of which referenced a source.
3   *Initiative In Energy, The Story of Dresser Industries*, Darwin Payne, 1979
4   *Engines—The Search for Power*, John Day, 1980
5   *The Gas Turbine*, Norman Davy, 1914
6   *Modern Gas Turbines*, Arthur W. Judge 1950

Chapter 2

General Compressor Theory

Compressors are mechanical devices used to increase the pressure of air, gas or vapor and in the process move it from one location to another. The inlet or suction pressure can range from low sub-atmospheric pressure levels to any pressure level compatible with piping and vessel strength limits. The ratio of absolute discharge pressure to absolute suction pressure is the compressor pressure ratio (*CR*—see Appendix B2, Glossary of Terms). Stage compression is limited to the mechanical capabilities of the compressor and, generally, approaches a *CR* of 4. To achieve high pressures multiple stages must be employed.

Compression theory is primarily defined by the Ideal Gas Laws and the First & Second Laws of Thermodynamics. As originally conceived the Ideal Gas Law is based on the behavior of pure substances and takes the following form:

$$Pv = RT \qquad (2\text{-}1)$$

Where

P = Absolute Pressure
v = Specific Volume
R = Gas Constant
T = Absolute Temperature

This equation is based on the laws of Charles, Boyle, Gay-Lussac and Avogadro (see Appendix B2 Glossary of Terms).

Note all properties should be defined in the same measuring system (for example either the English system or the metric system). Conversion factors listed in Appendix B3 can be used to assist in obtaining consistent units. Table 2-1 sums up the two systems.

Table 2-1.

Parameter	Symbol	English System	Metric System
Pressure	P	Absolute pressure (psia)	Pascals or Kilopascals
Temperature	T	Absolute temperature ($^\circ$R)	Degrees Kelvin ($^\circ$K)
Specific Volume	v	Cubic inches per pound	Cubic centimeters per gram or cubic meters per kilogram
Universal Gas Constant	$\bar{R}$	1545 ft-lbf/ lbm $^\circ$R	8.3144 kN m/ kmol $^\circ$K

The ideal gas law can be manipulated to obtain several useful relationships. By multiplying both sides of the equation by the mass "m" of the gas the specific volume becomes total volume:

$$V = mv$$

$$PV = mRT \tag{2-2}$$

Considering that the mass of any gas is defined as the number of moles times its molecular weight than (see Avogadro's Law in Appendix B2):

$$m = n \times mw$$

and

$$PV = n \times mw \times RT \tag{2-3}$$

and

$$\bar{R} = mw \times R \tag{2-4}$$

Where $\bar{R}$ is the universal gas constant

$$\frac{P_1 V_1}{T_1} = n\bar{R} = mR = \frac{P_2 V_2}{T_2} \tag{2-5}$$

The specific gas constant may be obtained using the universal gas constant and equation 2-4 above. However, Table 2-2 list the specific gas constant for some of the more common gases.

Table 2-2

Gas	Formula	Molecular Weight	Specific Gas Constant $\dfrac{ft \times lb_f}{lb_m} \times °R$
Helium	He	4.003	386.2
Carbon Monoxide	CO	28.01	55.18
Hydrogen	H_2	2.016	766.6
Nitrogen	N_2	28.02	55.16
Oxygen	O_2	32.00	48.29
Carbon Dioxide	CO_2	44.01	35.12
Sulfur Dioxide	SO_2	64.07	24.12
Water Vapor	H_2O	18.02	85.78
Methane	CH_4	16.04	96.35
Ethane	C_2H_6	30.07	51.40
Iso-butane	C_4H_{10}	58.12	26.59

Dividing both sides by "time" the total volume becomes volumetric flow and the mass flow per unit time becomes the mass flow rate "W."

$$PQ = WRT \qquad (2-6)$$

Where

Q = Volumetric Flow Rate
W = Mass Flow Rate

A pure substance is one that has a homogeneous and constant chemical composition throughout all phases (solid, liquid and gas). For most compressor applications a mixture of gases may be considered a pure substance as long as there is no change of phase. The significance of introducing this concept is that the state of a simple compressible pure substance is defined by two independent properties.

An additional term may be considered at this time to correct for deviations from the ideal gas laws. This term is the compressibility factor "Z."

Therefore, the ideal gas equation becomes

$$Pv = ZRT \qquad (2\text{-}7)$$

and

$$PQ = ZWRT \qquad (2\text{-}8)$$

Compressor performance is generally shown as pressure ratio plotted against flow. (Note: it is more accurate to use head instead of pressure ratio, because head takes into account the compressibility factor of the gas, molecular weight, temperature, and the ratio of specific heat of the gas—and corrected flow—all at constant speed). This is discussed in more detail later in this chapter.

Other relationships that are also useful are:

Reduced Temperature and Pressure

$$T_r = \frac{T}{T_c} \qquad (2\text{-}9)$$

$$P_r = \frac{P}{P_c} \qquad (2\text{-}10)$$

Where
  Tr = Reduced Temperature
  Pr = Reduced Pressure
  Tc = Critical Temperature
  Pc = Critical Pressure
  T = Observed Temperature
  P = Observed Pressure

Partial Pressure

The total pressure is equal to the sum of the partial pressures

$$P = P_1 + P_2 + P_3 + \ldots \qquad (2\text{-}11)$$

This relationship is defined by Dalton's Law (see Appendix B2).

If the total pressure of the mixture is known than the partial pressure can be calculated from the mole fraction.

$$P = P_1 + P_2 + P_3 + \dots \qquad (2\text{-}12)$$

The mole fraction "x" is

$$x_1 = \frac{M_1}{M_m} \; ; \; x_2 = \frac{M_2}{M_m} \; ; \; x_3 = \frac{M_3}{M_m} \; ; \qquad (2\text{-}13)$$

The partial pressure may then be calculated as follows

$$P_1 = x_1 \times P; \; P_2 = x_2 \times P; \; P_3 = x_3 \times P \qquad (2\text{-}14)$$

$$x_1 + x_2 + x_3 + = 1.0 \qquad (2\text{-}15)$$

First Law of Thermodynamics

The first law of thermodynamics states that energy cannot be created or destroyed but it can be changed from one form to another.

$$Q_h = \overline{W}_w + \Delta E \qquad (2\text{-}16)$$

Where

Q_h = Heat into the system
$\overline{W}_w$ = Work by the system
ΔE = Change in system energy

Note: the symbol $\overline{W}$ is used to denote work, whereas the symbol W indicates weight flow rate.

Second Law of Thermodynamics

The second law of thermodynamics states that the entropy of the universe always increases. This is the same law that indicates that perpetual motion machines are not possible.

$$\Delta s \geq 0 \qquad (2\text{-}17)$$

Horse Power Calculations

The brake horsepower (BHP) required to drive the compressor can be determined by calculating the gas horsepower (GHP) and then correcting for mechanical losses.

$$GHP = \frac{H_d \times W_g}{60 \times 33,000 \times E_p} \tag{2-18}$$

Where

$\quad$ BHP $\quad$ = Brake horsepower

$\quad$ H_d $\quad$ = Head (adiabatic) – ft-lb/lb

$\quad$ W_g $\quad$ = Weight flow of the gas – lbs/hr

$\quad$ E_p $\quad$ = Adiabatic efficiency

and

$$W_g = \frac{SCFD \times MW}{24 \times 379.5} \tag{2-19}$$

If capacity is available GHP can be calculated directly.

$$GHP = \frac{\left\{ Q_1 \times P_1 \times \left(\dfrac{Z_{av}}{Z_1}\right) \times \left(\dfrac{CR^{\frac{k-1}{k}}-1}{\frac{k-1}{k}} \right) \right\}}{(229 \times E_P)} \tag{2-20}$$

Where

$\quad$ Q_1 $\quad$ = Inlet cubic feet/minute (ICFM)

$\quad$ P_1 $\quad$ = Suction pressure—psia

$\quad$ CR $\quad$ = Compression ratio

$\quad$ K $\quad$ = Specific heat ratio

$\quad$ Zav $\quad$ = Average gas compressibility factor

$\quad$ Z_1 $\quad$ = Gas compressibility factor at compressor inlet

Then brake horsepower is

$$BHP = GHP \times (1 + \% \text{ Mechanical Losses}) \tag{2-21}$$

Each compressor type has its own unique characteristics that will be covered in the section specific to that compressor type.

Note: The physical and thermodynamic properties of many gases are provided in Appendix C, courtesy of the Gas Processors Suppliers Association (GPSA).

Chapter 3

Compressor Types

DYNAMIC COMPRESSORS

Two types of dynamic compressors are in use today—they are the *axial* compressor and the *centrifugal* compressor. The axial compressor is used primarily for medium and high horsepower applications, while the centrifugal compressor is utilized in low horsepower applications.

Both the axial and centrifugal compressors are limited in their range of operation by what is commonly called **stall** (or **surge)** and **stone wall**. The **stall** phenomena occurs at certain conditions of flow, pressure ratio, and speed (rpm), which result in the individual compressor airfoils going into stall similar to that experienced by an airplane wing at a high angle of attack. The stall margin is the area between the steady state operating line and the compressor stall line. Surge or stall will be discussed in detail later in this chapter. **Stone wall** occurs at high flows and low pressure. While it is difficult to detect. **Stone wall** is manifested by increasing gas temperature.

Axial Compressors

Gas flowing over the moving airfoil exerts lift and drag forces approximately perpendicular and parallel to the surface of the airfoil (Figure 3-1). The resultant of these forces can be resolved into two components:

1. the component parallel to the axis of the compressor represents an equal and opposite rearward force on the gas—causing an increase in pressure;

2. a component in the plane of rotation represents the torque required to drive the compressor.

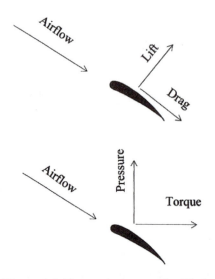

Figure 3-1. Forces Acting on the Blades

From the aerodynamic point of view there are two limiting factors to the successful operation of the compressor. They are the angle of attack of the airfoil and the speed of the airfoil relative to the approaching gas (Figure 3-2). If the angle of attack is too steep, the flow will not follow the concave surface of the airfoil. This will reduce lift and increase drag. If the angle of attack is too shallow, the flow will separate from the concave surface of the airfoil. This also results in increased drag.

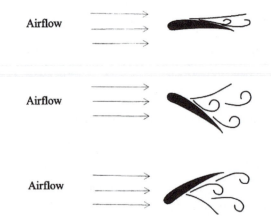

Figure 3-2. Airfoil Angle of attack Relative to Approaching Air or Gas

If the speed of the airfoil relative to the gas is too high, a shock will develop as the gas exceeds the speed of sound trying to accelerate as it passes around the airfoil. This shock will cause turbulent flow and result in an increase in drag. Depending on the length of the airfoil, this excessive speed could apply only to the tip of the compressor blade. Manufacturers have overcome this, in part, by decreasing the length of the airfoil and increasing the width (or chord).

For single-stage operation, the angle of attack depends on the relation of flow to speed. It can be shown that the velocity relative to the blade is composed of two components: the axial component depends on the flow velocity of the gas through the compressor, and the tangential component depends on the speed of rotation of the compressor (Figure 3-3). Therefore, if the flow for a given speed of rotation (rpm) is reduced, the direction of the gas approaching each blade is changed so as to increase the angle of attack. This results in more lift and pressure rise until the stall angle of attack is reached.

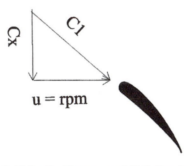

Figure 3-3. Velocity Component Relative to Airfoil

This effect can be seen on the compressor characteristic curve. The characteristic curve plots pressure against flow (Figure 3-4). The points on the curve mark the intersection of system resistance, pressure, and flow. (Note that opening the bleed valve reduces system resistance and moves the compressor operating point away from surge.) The top of each constant speed curve forms the loci for the compressor stall (surge) line.

Therefore, the overall performance of the compressor is depicted on the compressor performance map, which includes a family of constant speed (rpm) lines (Figure 3-5). The efficiency islands are included to show the effects of operating on and off the design point. At the design speed and flow, the angle of attack relative to the blades is optimum

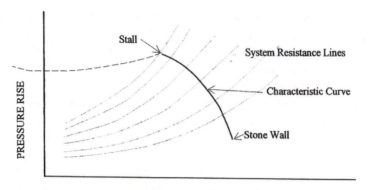

Figure 3-4. Compression System Curve

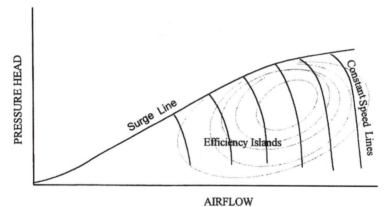

AIRFLOW

Figure 3-5. Compressor Performance Curve

and the compressor operates at peek efficiency. If flow is reduced at a constant speed, the angle of attack increases until the compressor airfoil goes into stall.

As flow is increased at a constant speed the compressor characteristic curve approaches an area referred to as "stone wall." Stone wall does not have the dynamic impact that is prevalent with stall, but it is a very inefficient region. Furthermore, operation at or near stonewall will result in overtemperature conditions in the downstream process.

From the mechanical point of view, blade stresses and blade vibration are limiting factors. The airfoil must be designed to handle the varying loads due to centrifugal forces, and the load of compressing

gas to higher and higher pressure ratios. These are conflicting requirements. Thin, light blade designs result in low centrifugal forces, but are
limited in their compression-load carrying ability; while thick, heavy
designs have high compression-load carrying capability, but are limited
in the centrifugal forces they can withstand. Blade vibration is just as
complex. There are three categories of blade vibration: resonance, flutter,
and rotating stall. They are explained here.

> *Resonance*—As a cantilever beam, an airfoil has a natural frequency
> of vibration. A fluctuation in loading on the airfoil at a frequency
> that coincides with the natural frequency will result in fatigue
> failure of the airfoil.

> *Flutter*—A self-excited vibration usually initiated by the airfoil ap-
> proaching stall.

> *Rotating Stall*—As each blade row approaches its stall limit, it does
> not stall instantly or completely, but rather stalled cells are formed
> (see Rotating Stall Figure 6-1). Stall is discussed in more detail in
> Chapter 6.

The best way to illustrate flow through a compressor stage is by
constructing velocity triangles (Figure 3-6). Gas leaves the stator vanes
at an absolute velocity of C_1 and direction θ_1. The velocity of this gas
relative to the rotating blade is W_1 at the direction β_1. Gas leaves the
rotating stage with an absolute velocity C_2 and direction θ_2, and a relative velocity W_2 and direction β_2. Gas leaving the second stator stage has
the same velocity triangle as the gas leaving the first stator stage. The
projection of the velocities in the axial direction are identified as Cx, and
the tangential components are Cu. The flow velocity is represented by
the length of the vector. Velocity triangles will differ at the blade hub,
mid-span, and tip just as the tangential velocities differ.

Pressure rise across each stage is a function of the gas density, ρ,
and the change in velocity. The pressure rise per stage is determined
from the velocity triangles:

$$\Delta P = \frac{\rho}{2g_C} (W_1^2 - W_2^2) + \frac{\rho}{2g_C} (C_2^2 - C_3^2) \qquad (3\text{-}1)$$

This expression can be further simplified by combining the differential
pressure and density, and referring to feet of head.

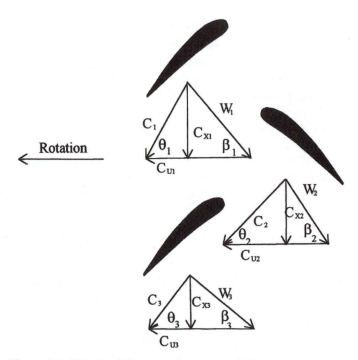

Figure 3-6. Velocity Diagrams for an Axial Flow Compressor.

$$\text{Head} = \frac{\mu}{g_C} \times \Delta C = \frac{\Delta P}{\rho} \qquad (3\text{-}2)$$

where $\Delta P/\rho$ is the pressure rise across the stage and head is the pressure rise of the stage measured in feet head of the fluid flowing. The standard equation for compressor head is given below.

$$\text{Head} = \frac{Z_{ave}T_s}{MW}\left(\frac{R_c^\sigma - 1}{\sigma}\right)$$

$$(3\text{-}3)$$

where Z_{ave} is the average compressibility factor of gas, and MW is the mole weight.

Before proceeding further, we will define the elements of an airfoil (Figure 3-7).

Fixed Blade Row   Turning Angle $\equiv \varepsilon = \beta_2' - \beta_1'\text{xx}$
Moving Blade Row   Chord Length

Camber Line

Camber or Blade Angle $\equiv \beta_2 - \beta_1$

Inlet Blade Angle $\equiv \beta_1$

Exit Blade Angle $\equiv \beta_2$

Inlet Flow Angle $\equiv \beta_1'$

Exit Flow Angle $\equiv \beta_2'$

Angle of Deviation $\equiv \beta_2 - \beta_2'$

Stagger Angle $\equiv \gamma$

Pitch $\equiv s$

Leading Edge

Trailing Edge

Maximum Thickness $\equiv t$

Width $\equiv w$

Height $\equiv h$

Aspect Ratio $\equiv$ ratio of blade height
to blade chord

Angle of Attack or angle if
incidence $\equiv i = \beta_1 - \beta_1'$

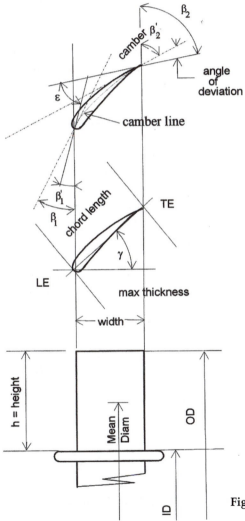

Figure 3-7. Elements of an Airfoil.

Centrifugal Compressors

The centrifugal compressor, like the axial compressor, is a dynamic machine that achieves compression by applying inertial forces to the gas (acceleration, deceleration, turning) by means of rotating impellers.

The centrifugal compressor is made up of one or more stages, each stage consisting of an impeller and a diffuser. The impeller is the rotating element and the diffuser is the stationary element. The impeller consist of a backing plate or disc with radial vanes attached to the disc from the hub to the outer rim. Impellers may be either open, semi-enclosed, or enclosed design. In the open impeller the radial vanes attach directly to the hub. In this type of design the vanes and hub may be machined from one solid forging, or the vanes can be machined separately and welded to the hub. In the enclosed design, the vanes are sandwiched between two discs. Obviously, the open design has to deal with gas leakage between the moving vanes and the non-moving diaphragm, whereas the enclosed design does not have this problem. However, the enclosed design is more difficult and costly to manufacture.

Generally gas enters the compressor perpendicular to the axis and turns in the impeller inlet (eye) to flow through the impeller. The flow through the impeller than takes place in one or more planes perpendicular to the axis or shaft of the machine. This is easier to understand when viewing the velocity diagrams for a centrifugal compressor stage. Although the information presented is the same, Figure 3-8 demonstrates two methods of preparing velocity diagrams.

Centrifugal force, applied in this way, is significant in the development of pressure. Upon exiting the impeller, the gas moves into the diffuser (flow decelerator). The deceleration of flow or "diffuser action"

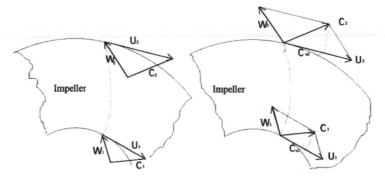

Figure 3-8. Velocity Diagrams for a Centrifugal Compressor

causes pressure build-up in the centrifugal compressor. The impeller is the only means of adding energy to the gas and all the work on the gas is done by these elements. The stationary components, such as guide vanes and diffusers, can only convert velocity energy into pressure energy (and incur losses). Pressure from the impeller eye to the impeller outlet is represented by the following:

$$P_m = \frac{G}{2g_c}[(C_2{}^2 - C_1{}^2) + (U_2{}^2 - U_1{}^2) + (W_1{}^2 - W_2{}^2)] = \frac{G}{2g_c}(C_{2u}U_2 - C_{1u}U_1) \quad (3\text{-}4)$$

The term $(C_2{}^2 - C_1{}^2)/2g_c$ represents the increase in kinetic energy contributed to the gas by the impeller. The absolute velocity C_1 (entering the impeller) increases in magnitude to C_2 (leaving the impeller). The increase in kinetic energy of the gas stream in the impeller does not contribute to the pressure increase in the impeller. However, the kinetic energy does convert to a pressure increase in the diffuser section. Depending on impeller design, pressure rise can occur in the impeller in relation to the terms $(U_2{}^2 - U_1{}^2)/2g_c$ and $(W_1{}^2 - W_2{}^2)/2g_c$. The term $(U_2{}^2 - U_1{}^2)/2g_c$ measures the pressure rise associated with the radial/centrifugal field, and the term $(W_1{}^2 - W_2{}^2)/2g_c$ is associated with the relative velocity of the gas entering and exiting the impeller. The ideal head is defined by the following relationship:

$$\text{Head}_{ideal} = P_m = \frac{1}{g_c} (C_2{}^2 - C_1{}^2)/2g_c \qquad (3\text{-}5)$$

$$\text{Flow coefficient } \phi \text{ is defined as} = \frac{Q}{AU} \qquad (3\text{-}6)$$

where
$$Q = \text{Cubic feet per second (CFS)}$$
$$A = \text{Ft}^2$$
$$U = \text{Feet per sec}$$

rewriting this equation $\qquad \phi = \dfrac{700Q}{D^3N} \qquad (3\text{-}7)$

where D is diameter, and N is rotor speed.

The flow coefficients are used in designing and sizing compressors and in estimating head and flow changes resulting as a function of tip speed (independent of compressor size or rpm). Considering a constant geometry compressor, operating at a constant rpm, tip speed is also constant. Therefore, any changes in either coefficient will be directly related to changes in head or flow. Changes in head or flow under these conditions result from dirty or damaged compressor impellers (or airfoils). This is one of the diagnostic tools used in defining machine health.

The thermodynamic laws underlying the compression of gases are the same for all compressors—axial and centrifugal. However, each type exhibits different operating characteristics (Figure 3-9). Specifically, the constant speed characteristic curve for compressor pressure ratio relative to flow is flatter for centrifugal compressors than for axial compressors. Therefore, when the flow volume is decreased (from the design point) in a centrifugal compressor, a greater reduction in flow is possible before the surge line is reached. Also, the centrifugal compressor is stable over a greater flow range than the axial compressor, and compressor efficiency changes are smaller at off design points.

For the same compressor radius and rotational speed the pressure rise per stage is less in an axial compressor than in a centrifugal compressor. But, when operating within their normal design range, the efficiency of an axial compressor is greater than a centrifugal compressor.

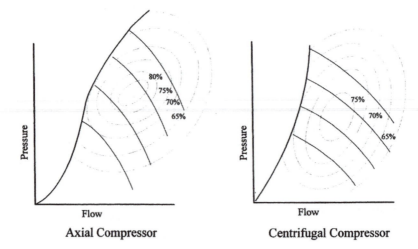

Figure 3-9. Comparing Characteristic Curves of Axial and Centrifugal Compressors.

Variations in Compressor Design

Variable Guide Vane Compressor

Variable guide vanes (VGVs) are used to optimize compressor performance by varying the geometry of the compressor. This changes the compressor characteristic curve and the shape and location of the surge line. In this way the compressor envelope can cover a much wider range of pressure and flow. In centrifugal compressors the variable guide vane (that is, the variable inlet vane) changes the angle of the gas flow into the eye of the first impeller. In the axial compressor up to half of the axial compressor stages may incorporate variable guide vanes. In this way the angle of attack of the gas leaving each rotating stage is optimized for the rotor speed and gas flow.

Variable guide vane technology enables the designers to apply the best design features in the compressor for maximum pressure ratio & flow. By applying VGV techniques the designer can change the compressor characteristics at starting, low-to-intermediate and maximum flow conditions. Thereby maintaining surge margin throughout the operating range. Thus creating the best of all worlds. The compressor map in Figure 3-10 demonstrates how the surge line changes with changes in vane angle.

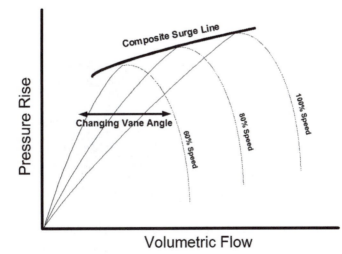

Figure 3-10. Composite Surge Limit Line Resulting From Variations In Vane Position

Variable guide vanes have an effect on axial and centrifugal compressors similar to speed: the characteristic VGV curves of the axial compressor being steeper than the characteristic VGV curves of the centrifugal compressor. The steeper the characteristic curve the easier it is to control against surge whereas the flatter characteristic curve the easier it is to maintain constant pressure control (or pressure ratio control).

POSITIVE DISPLACEMENT COMPRESSORS

There are many types of positive displacement compressors, but only the three major types will be discussed in this book. These are the blower, reciprocating compressor, and screw compressor. Blowers and screw compressors share some common features and a common history, while reciprocating compressors are noticeably different.

Blowers

The blower is a positive displacement compressor that was invented and patented in 1860 by Philander H. Roots and Francis M. Roots two brothers from Connersville, Indiana. Initially it was intended as a gas pump for use in blast furnaces. The blower design consists of two figure-eight elements or lobes. These lobes are geared to drive in opposite directions (see Figure 3-11).

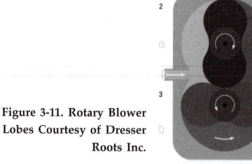

Figure 3-11. Rotary Blower Lobes Courtesy of Dresser Roots Inc.

Since the rotary lobes need to maintain clearance between each of the lobes a single stage blower can achieve only limited pressure ratio differential—typically 2.0. However, it is capable of compressing large volumes of gas at efficiencies up to 70%. Therefore, blowers are frequently used as boosters in various compression applications.

A typical two-lobe blower operating sequence is shown in Figure 3-12. Note that the upper rotor or lobe is turning clockwise while the lower lobe is turning counterclockwise.

Position #1: gas enters the lower lobe cavity from the left as compressed gas is being discharged from this cavity to the right and simultaneously gas is being compressed in the upper lobe cavity.

Position #2: the upper lobe cavity is about to discharge it's compressed gas into the discharge line.

Position #3: gas enters the upper lobe cavity from the left as compressed gas is being discharged from this cavity to the right and simultaneously gas is being compressed in the lower lobe cavity.

Position #4: the lower lobe cavity is about to discharge it's compressed gas into the discharge line.

Figure 3-13 shows a cross-section of a two-lobe blower. The blower discharges gas in discrete pulses, similar to the reciprocating compressor (pulsation control is discussed in more detail in the section on reciprocating compressors). These pulses may be transmitted to the downstream equipment or process and must be treated accordingly. For example, if the downstream piece of equipment is a reciprocating compressor, the scrubber and suction pulsation bottle will dampen the pulses from the blower. Some blower designs incorporate pulsation chambers in the body of the blower casing to help minimize pulses.

Blowers are used to boost low pressure gas (at or near atmospheric pressure) to the inlet of larger compressors. The various applications for blowers are listed in Table 3-1.

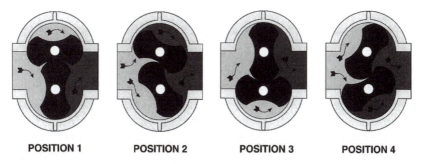

POSITION 1   **POSITION 2**   **POSITION 3**   **POSITION 4**

Figure 3-12. Typical Blower Operating Sequence

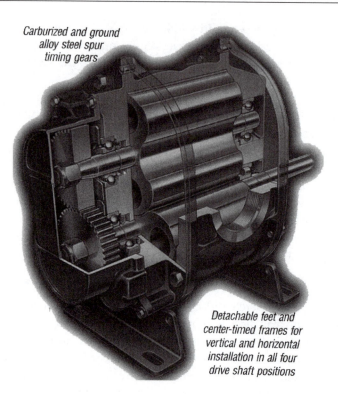

Carburized and ground
alloy steel spur
timing gears

Detachable feet and
center-timed frames for
vertical and horizontal
installation in all four
drive shaft positions

Figure 3-13. Roots Blower Courtesy of Dresser-Roots, Inc.

Table 3-1. Blower Applications

Gas Boosting
Landfill Extraction
Digester Gas
Process Vacuum
Gas Recycling
Direct Reduced Iron
Gas Blowing

Blowers are usually belt driven with an electric motor as the driver. Chain driven applications are also used where the driver is close-coupled to the blower as in reciprocating engine/turbocharger applications. An exploded view of a two-lobe blower is shown in Figure 3-14.

Table 3-2 list the general causes of problems encountered in blower operation.

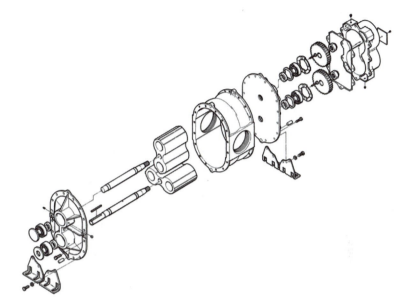

Figure 3-14. Exploded View of the Roots Universal RAI Blower Courtesy of Dresser Roots, Inc.

Most problems can be avoided by first checking that the:
- Driver operates properly before connecting it to the blower;
- Blower turns freely before connecting it to the driver and process piping
- Oil type is correct and
- Oil reservoir is filled to the proper level

Blowers have been developed in both two-lobe and three-lobe configurations. A three-lobe rotor is shown in Figure 3-15.

Reciprocating Compressors

The reciprocating, or piston compressor, is a positive displacement compressor that uses the movement of a piston within a cylinder to move gas from one pressure level to another (higher) pressure level. The simplest example of this is the bicycle pump used to inflate a bicycle tire.

In the bicycle pump example we have the piston attached to a long rod, a pump cylinder (a tube or pipe closed on each end), two ball-type check valves (one inlet and one outlet), a flex hose and a connection de-

Table 3-2. Troubleshooting Chart

TROUBLE	POSSIBLE CAUSE
Will Not Start	Check Driver Uncoupled Impeller Stuck Dirty blower elements
No flow	Wrong rotation Obstruction in inlet or discharge piping
Capacity Low	Speed too low Belt tension too loose Compressor ratio too high Obstruction in inlet or discharge piping Dirty blower elements
Excessive Power Draw (Motor Amp High)	Compressor ratio too high Obstruction in inlet or discharge piping Dirty blower elements
Vibration	Misalignment Impellers unbalanced Impellers contacting Driver or blower casings loose Foundation not level Worn bearings/gears Piping pulsations Dirty blower elements
Bearing/Gear Damage	Coupling/belt misaligned Belt tension excessive
Breather Blow-by Excessive	Broken seals or O-ring Oil level too high Oil type incorrect

vice. The piston is as simple as a leather or rubber disk with two smaller diameter metal disks on either side to give the piston strength and stiffness. The cylinder is the tube that the leather-sandwiched piston moves within. The inlet or suction spring-loaded ball valves allow ambient air to flow into the cylinder as the piston is partially withdrawn from the cylin-

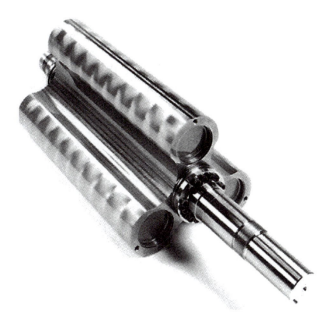

Figure 3-15. Three-lobe Blower Rotor Courtesy of Dresser Roots, Inc.

der (creating a partial vacuum) and closes while the piston compresses the air within the cylinder on the compression stroke; while the discharge ball valve (which is closed on the suction cycle) opens to let the compressed air flow out of the cylinder and into the receiver or tire.

Reciprocating compressors are used in many different industries:
- Oil & gas
  - Oil refineries
  - Gas gathering
  - Gas processing
- Transportation
  - Gas pipelines
- Chemical plants
- Refrigeration plants (large and small)

Reciprocating compressors types include the:
- Single-cylinder compressor
- Multi-cylinder balanced opposed compressor, and the
- Integral-engine compressor.

Single-cylinder Reciprocating Compressor

The single-cylinder reciprocating compressor (Figure 3-16) contains all the elements of any of the reciprocating compressors types.

Major elements of this compressor are the:

— Frame
— Piston
— Cylinder
— Crankshaft
— Connecting Rod
— Crosshead
— Piston Rod

The **frame, crankshaft, connecting rod** and **piston rod** define the strength or horsepower capability of the compressor. This overall strength or horsepower capability is combined into one measurement and that is rod load. Rod load is quoted in the English measurement system as pounds and in the metric system as Newtons.

Figure 3-16. Worthington HB GG 24 x 13 Single Stage Compressor before overhaul. Courtesy of Sun Engineering Services, Inc. and Power & Compression Systems Co.

Rod load is the algebraic sum of the inertia forces acting on the compressor frame, crankshaft, connecting rod, crosshead, piston and piston rod and gas loads. Therefore, rod load must take into account inertia forces of the reciprocating components and the gas load calculated at crank angle increments (usually 10 degree increments). Rod loads are defined as combined or overall loads, compression load and tension load.

$$\text{Compression Rod Load} = 0.785 \times [D_P^2 \times (P_D - P_S) + D_R^2 \times P_S] \qquad (3\text{-}8)$$

$$\text{Tension Rod Load} = 0.785 \times [D_P^2 \times (P_D - P_S) - D_R^2 \times P_S] \qquad (3\text{-}9)$$

Where:
D_P = Piston diameter – inches
D_R = Rod diameter – inches
P_D = Discharge pressure – psia
P_S = Suction pressure – psia

A change in sign of the combined or overall loads indicates a reversal in the vector loading on the bearing. This is discussed later in the section on bearings.

Frame

The frame or crankcase is generally made from cast iron although in some instances it is fabricated from steel plate. The crankcase forms the base upon which the crankshaft is supported. The crankcase also serves as the oil reservoir.

Crankshaft

The crankshaft, typically made of forged steel, consists of crankpins and bearing journals (Figure 3-17). While small crankshafts have been machined from a single forging, most crankshafts are fabricated from three parts: bearing journals, crankpins and interconnection pieces.

The crankshaft converts rotating motion to reciprocating linear motion via the connecting rod. The crankshaft journal connects to the journal bearing on the crank-end of the connecting rod and the crankpin connects to the journal bearing on the crosshead-end of the connecting rod.

Crosshead

The crosshead (Figure 3-18) rides in the crosshead guide moving linearly in alternate directions with each rotation of the crankshaft. The

Figure 3-17. Superior MH/HW Compressor Crankshaft Made From Either Forged Or Billet Heat Treated SAE 4140 Steel Courtesy of Cooper Cameron Corporation.

piston rod connects the crosshead to the piston. Therefore, with each rotation of the crankshaft the piston moves linearly in alternating directions.

The crosshead sits in the crosshead guide with the crankshaft on one end and the piston rod on the other. The crosshead guide may be an integral part of the "distance piece" or a separate section. The guide consists of two bearing surfaces: one on top of the guide and one on the bottom. As the crosshead moves back and forth within the guide

Figure 3-18. Crosshead Assembly (Piston Rod Not Shown) Courtesy of ACI Services, Inc.

the piston rod exerts a vertical motion on it. As a result the crosshead is alternately in contact with the bottom bearing surface and the top bearing surface. Care must be taken that piston rod droop is within acceptable limits so as not to result in the crosshead cocking excessively.

Note: See Chapter 12 Case History for the article by Gordon Ruoff on piston rod run-out.

Distance Piece

The distance piece is used to provide sufficient space to facilitate separation of toxic vapors from the oil system and the atmosphere. The "distance piece" mounts to the compressor frame. Besides housing the crosshead guide the distance piece also houses the oil wiper packing (Figure 3-19). The piston rod traverses through the packing as it moves from the head-end to the crank-end thereby acting to seal the process gas in the cylinder from the lubricating environment. Due to the wear created on the piston rod resulting from contact with the packing the piston is often hardened in this contact area. The rod packing is originally assembled when the piston and rod assembly is installed. However, after assembly the sealing rings can be removed and replaced as needed.

Piston Rod

The piston rod is threaded on both ends and may even have a collar on the end that connects to the piston (Figure 3-20). Connecting the piston rod to the piston is an elaborate procedure. The piston must be secured to the rod so that the alternating inertial loading created

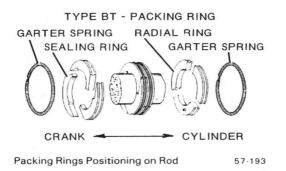

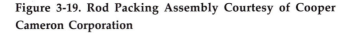

Packing Rings Positioning on Rod   57-193

Figure 3-19. Rod Packing Assembly Courtesy of Cooper Cameron Corporation

Figure 3-20. Piston, Piston Rod, Rings and Wear Band Courtesy of ACI Services, Inc.

by the accelerating, decelerating and stopping of the piston with each revolution of the crankshaft does not put undue stresses on the rod or the piston ends. This is accomplished by stretching the rod between the crank-end collar and the head-end piston rod nut while simultaneously compressing the piston ends. The sketch below, though grossly out of proportion, illustrates the process (Figure 3-21).

Compressor manufacturers expend considerable amount of engineering and test man-hours to arrive at the proper lubricant and torque

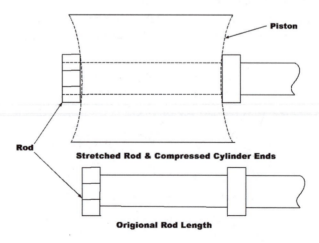

Figure 3-21. Exaggerated Representation of Piston Deformation and Rod Stretch.

to stretch the rod and compress the piston while staying within the elastic limits of the materials of both parts.

The crank-end of the piston rod is threaded into the crosshead and a lock nut is used to guard against the rod backing out of the crosshead. The locknut also functions as a counterweight to compensate for loading on opposite sides of the crankshaft.

Piston

Pistons may be single-acting or double-acting. Single-acting pistons compress the process gas as the pistons moves towards the head-end of the cylinder only (see Figure 3-24). Double-acting pistons compress the process gas as the piston moves in both directions (see Figure 3-23).

Also pistons may be cast from one piece or assembled from several. Although piston balance has not been addressed by most manufacturers there have been cases where piston imbalance has caused rod failure.

Cylinders

There are two types of compressor cylinder designs: "Valves In Bore" and "Valves Out of Bore." The valves in bore design has the compressor valves located radially around the cylinder bore within the length of the cylinder bore. These cylinders have the highest percentage of clearance due to the need for scallop cuts at the head-end and crank-

Figure 3-22. Three Piece Cast Iron Pistons With TFE Riders And Flanged Rods Courtesy of ACI Services, Inc.

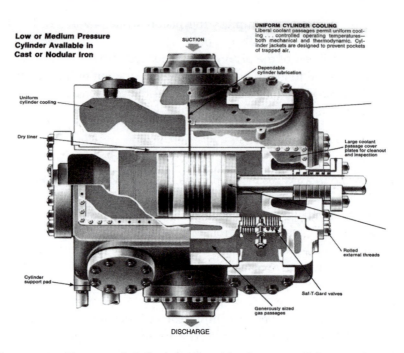

Figure 3-23. Piston and Cylinder Assembly Courtesy of Cooper Cameron Corporation

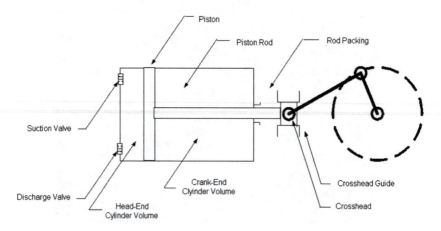

Figure 3-24. Single Acting Piston Operation

end of the cylinder bore to allow entry and discharge for the process gas.

The valves out of bore design consists of compressor valves at each end of the cylinder. While this design provides a lower percent clearance it is more maintenance intense. The unusually large outer head requires a large number of bolts that must be removed for routine service of the piston, wear band and compression rings. Sealing gas pressure on the crank-end is also problematic.

A compressor cylinder houses the piston, suction & discharge valves, cooling water passages (or cooling fins), lubricating oil supply fittings and various unloading devices. Medium to large compressor cylinders also include a cylinder liner.

The volume change during each stroke of the piston is the piston displacement (see Appendix F Compressor Cylinder Displacement Curves. The minimum cylinder volume is the clearance volume. There must be some clearance volume in a cylinder to avoid contact with the head-end or crank-end of the piston. See "Sizing & Selection" at the end of the Reciprocating Compression Section for details on calculating volumetric clearance and brake horsepower.

While a cylinder's capacity is physically fixed, several methods can be employed to reduce capacity and/or pressure differential.

Volume pockets (fixed and variable), volume bottles, suction valve unloaders and speed can be used to vary the compressor throughput. These will be discussed in more detail in the section on controls.

Valves

As discussed in the bicycle pump example, valves are used to control the gas flow into and out of the cylinder. There are suction valves and discharge valves.

The suction valves open at the start of the suction stroke to allow gas to be drawn into the cylinder and close at the start of the compression stroke.

The discharge valves open when sufficient pressure builds up in the cylinder to overcome the combination of valve plate(s) spring load and downstream pressure. At the end of the compression stroke (just prior to the piston reversing direction) the pressure across the valve equalizes and valve plate(s) spring load closes the discharge valve. Valve plate (or poppets) should be smooth and should not bounce as these shorten valve life. Figure 3-25 shows the correct valve operation. Figure

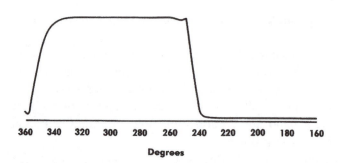

360   340   320   300   280   260   240   220   200   180   160

Degrees

Figure 3-25. Correct Valve Operation Courtesy of Roger Vaglia, Product Manager (Retired), White Superior Compressors 1962-2000

3-26 indicates an incorrect valve operation, in this case resulting from higher than necessary spring force.

Valve designs may employ either plates or poppets and there are many variations on these designs.

The pros and cons in the use of each are dependent on the compression ratio and the gas characteristics. Figures 3-27, 3-28 and 3-29 provide a general overview of the two valve types.

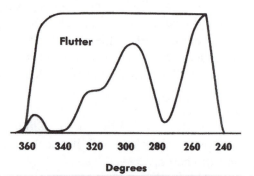

Figure 3-26. Incorrect Valve Operation—Valve Overspringing Courtesy of Roger Vaglia, Product Manager (Retired), White Superior Compressors 1962-2000

Figure 3-27 Compressor Poppet Valve Courtesy of Cooper Cameron Corporation

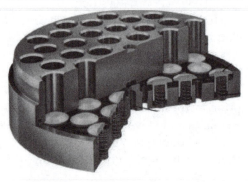

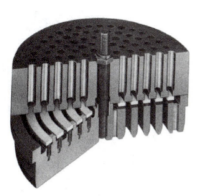 

Figure 3-28 Compressor Plate
Valve Courtesy of Cooper Cameron Corporation

Figure 3-29. Poppet Valve Poppets Courtesy of ACI Services,
Inc.

Suction valves are installed in the cylinder valve cavity above the cylinder centerline and discharge valves are installed in the cylinder valve cavity below the cylinder centerline.

API Standard 618 "Reciprocating Compressors For Petroleum, Chemical, And Gas Industry Services" stipulates that the suction and discharge valves be designed and manufactured such that they cannot be inadvertently interchanged.

Suction valve unloaders (plug or finger type—see Figures 3-30, through 3-33 and volume pockets Figure 3-34) can be used to regulate compressor throughput. The suction valve unloaders prevent gas pressure from building up in the cylinder thereby reducing that cylinder's output.

A drawback of this approach is that cylinder gas temperature will rise if the same cylinder end is unloaded indefinitely. To prevent this from happening the control system must monitor cylinder gas temperatures and cycle through various cylinder loading configurations (this is addressed in more detail in Chapter 5, Throughput Control).

An alternate approach is the Head-End Bypass offered by ACI Services, Inc. Utilizing SimPlex™, the configuration the head-end is unloaded by porting the head-end cylinder compressed gas directly back to suction. The DuPlex™ adds a pocket unloader to increase the head-

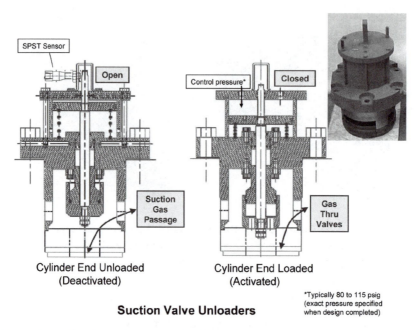

Cylinder End Unloaded
(Deactivated)

Cylinder End Loaded
(Activated)

Suction Valve Unloaders

*Typically 80 to 115 psig
(exact pressure specified
when design completed)

Figure 3-30. Plug-Type Suction Valve Unloader Courtesy of ACI Services, Inc.

end clearance for a partial reduction in throughput and the head-end unloader for maximum head-end unloading.

An advantage of this approach is that the head-end may be unloaded indefinitely. As with the suction valve unloader approach additional control logic must be incorporated into the compressor controller to activate the unloader.

The plug type unloader (Figure 3-30) uses a plug or piston through the center of the valve. When the plug is retracted gas pressure cannot build up in the cylinder.

The finger type unloader uses a set of "fingers" on each valve plate to hold the valve plates open thus preventing gas pressure build up in the cylinder (see Figure 3-32 and 3-33). There are some drawbacks with each type of unloader:

• The fingers, in the finger-type unloader, are subject to fatigue due to repeated use and could break off and be ingested into the cylinder.

• Liquids and contaminates in the gas could cause the plug in the plug-type unloaded to stick or seize.

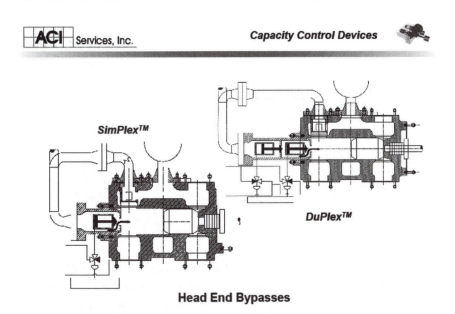

Figure 3-31. and Head-end Unloaders Courtesy of ACI Services, Inc.

Figure 3-32. Finger-type Unloader Courtesy of Cooper Cameron Corporation

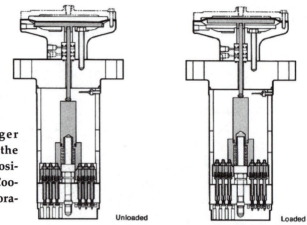

Figure 3-33. Finger Type Unloader in the Open and Closed Positions Courtesy of Cooper Cameron Corporation

Volume pockets increase the clearance volume (see the discussion on clearance volume and volumetric efficiency later in this chapter). This is another way to vary compressor throughput. As seen in Figure 3-34 retracting the plug increases the cylinder volume by the amount of the bottle capacity.

The only limitation to the size or number of volume bottles that can be added to a cylinder is the physical space available.

Lubrication

Lubrication is required for the main bearings, the crankshaft bearings, the crankpin bearings, the crosshead pin and the crosshead guide. Lubrication is also required for the piston.

Several methods are available to provide lubrication to the bearings and crosshead:

* Forced-feed method
* Passive gravity-feed method

The forced-feed oil lubrication uses a pump to move the oil through the compressor oil feed passages. A forced feed lube oil system is depicted in Figure 3-35.

The passive gravity method, as shown in Figure 3-36, uses a scoop or oil collector to move oil from the reservoir to a high point in the compressor. From that location the oil flows due to gravity to the bearings throughout the compressor.

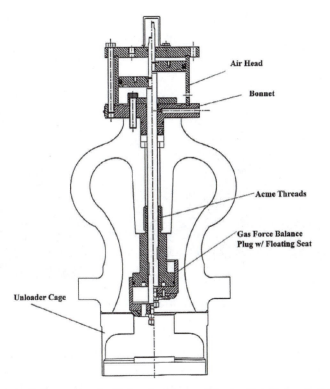

Figure 3-34. Volume Bottle Unloader Assembly in the Unloaded Position Courtesy of ACI Services, Inc.

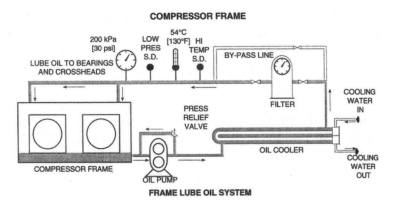

Figure 3-35. Compressor Frame Lubrication Oil System Courtesy of Petroleum Learning Programs, 305 Wells Fargo Dr., Houston, Texas 77090.

Figure 3-36. Oil Scoop can be seen directly below the bearing cap nut. Courtesy of Power & Compression systems

The force-feed lubrication system employs a shaft driven pump as the primary oil pump and an electric motor driven pump as the backup pump. Also included are dual lubrication filters. The force-feed lubrication method is usually employed in higher speed compressors (over 600rpm).

In cooler (or cold) climates a lubrication oil heater is installed in the reservoir. This oil heater is used to heat oil to a temperature at which oil will flow freely. For mineral oil this is approximately 35°F.

A separate lubrication system is employed to lubricate the piston within the cylinder. Since this method adds oil, an incompressible fluid, into the cylinder care must be taken not to add too much oil. Excessive oil supply could accumulate in the cylinder resulting in damage to the rod, piston or cylinder. The manufacturer will normally provide instructions to set the correct amount of oil supplied to the cylinder. Similar information is provided in the charts in Appendix G—Compressor Cylinder Lubrication.

Bearings

Hydrodynamic bearings are used almost exclusively in reciprocating compressors. The older, heavy-duty compressors use the split bear-

Figure 3-37. Connect-
ing Rod and Bearings
Courtesy of ACI Ser-
vices. Inc.

ing design. The newer, lighter compressors use the tilt-pad designs.

Due to the way the oil distribution paths are machined into the bearing holders and supports the crankshaft may only be rotated in one direction to maintain proper oil supply.

Bearing life is affected by the condition and temperature of the lubricating oil, rod load, and rod reversal. The rod load should change from compression to tension and back to compression with each revolution. That is with each revolution the load vector on the bearing must change to allow continuous lubrication to all bearing surfaces. If the load vector does not change the flow of lube oil becomes inadequate eventually leading to bearing failure. Rod reversal is defined as the minimum distance, measured in degrees of crank revolution, between each change in sign of the force in the combined rod-load curve (Figure 3-38).

Rod reversal less than 15 degrees crank angle is cause for concern.

Pulsation Bottles

While strictly speaking pulsations bottles are not an integral part of the reciprocating compressor, they are a necessary part of the recip-rocating compressor package. Pulsation bottles are placed at the inlet to the compressor and at the discharge from each compression stage. The pulsation bottle's stated purpose is to dampen pressure waves created by each compression pulse.

The suction pulsation bottle dampens the pressure pulse to the suction piping. Similarly the discharge pulsation bottles dampen the pressure pulses to the discharge piping. These pulses, if transmitted to

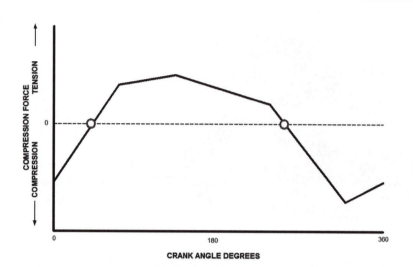

Figure 3-38. Red Reversal is the Shortest Distance, in Crank Angle Degrees, from Point to Point.

either the suction piping or discharge piping could be very detrimental to the operation and safety of the compressor.

Suction bottles are mounted on top of the compressor and discharge bottles are mounted beneath the compressor cylinders (Figure 3-39).

Sizing suction and discharge pulsation bottles is a complicated process. Although there are software programs that help with the calculations they should only be used by engineering personnel with experience in the seizing process and the software program. The American Petroleum Institute (API) suggests three design approaches to be followed when seizing pulsation bottles. The basic API guidelines are repeated here for the reader's convenience. However, it is strongly suggested that the entire specification be read and applied.

Reference Paragraph 3.9.2.1

"Design Approach 1: Pulsation control through the use of pulsation suppression devices designed using proprietary and/or empirical techniques to meet pulsation levels required in 3.9.2.5 based on the normal operating condition."

"Design Approach 2: Pulsation control through the use of pulsation suppression devices and proven acoustical simulation techniques in conjunc-

Figure 3-39. Dresser-Rand 7 HOSS-4 Compressors with Suction & Discharge Pulsation Bottles Courtesy of Power & Compression Systems

tion with mechanical analysis of pipe runs and anchoring systems (clamp design and spacing) to achieve control of vibrational response."

"Design Approach 3: The same as Design Approach 2, but also employing a mechanical analysis of the compressor cylinder, compressor manifold, and associated piping systems including interaction between acoustical and mechanical system responses as specified in 3.9.2.6"*

In order to size the suction and discharged bottles accurately detailed, information about the compressor and the suction and discharge piping must be available. This study is referred to as the Harmonic Resonance Analysis. Failure to perform the harmonic resonance analysis could result in pipe vibrations severe enough to rupture the gas piping.

The isometric piping drawing (see Figure 3-40) describes the piping layout, changes in elevation, size and changes in size. This information is critical to the safe design and operation of the compressor package.

---

*API Standard 618 "Reciprocating Compressors For Petroleum, Chemical, And Gas Industry Services." Latest Edition.

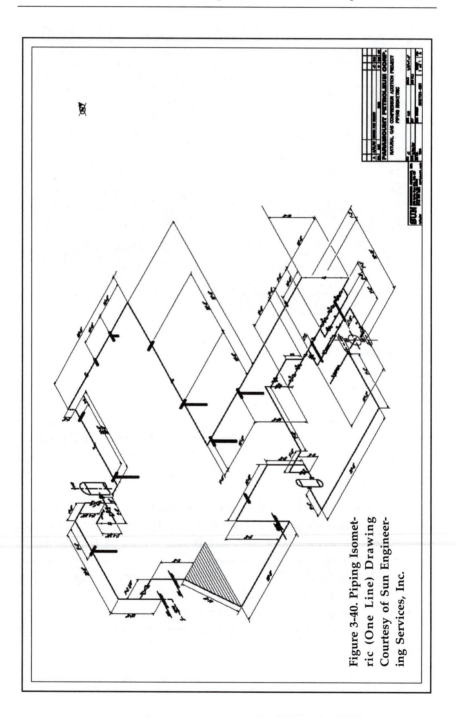

Figure 3-40. Piping Isometric (One Line) Drawing Courtesy of Sun Engineering Services, Inc.

Figure 3-41 is the overhauled and upgraded compressor shown in Figure 3-16. These two pictures demonstrate the longevity and usefulness of compression equipment.

Balanced Opposed Reciprocating Compressor

The balanced opposed reciprocating compressor as shown in Figure 3-42 makes use of multiple cylinders (as well as a crankshaft with multiple crankpin throws, connecting rods, crossheads, pistons and piston rods). Everything that was discussed in the section on single-cylinder compressors applies to the balanced opposed compressor.

The only limiting factor to the number of cylinders installed on a balanced opposed compressor is the load capability of the frame, the ability to maintain main bearing alignment and speed.

Balanced opposed compressors up to 12 cylinders are not unusual. Generally speaking compressors with two to six cylinders operate at speeds up to 1200 rpm. Compressors with eight or more cylinders run at speeds of 300 rpm.

Figure 3-41. Worthington HB GG 24 x 13 Single-stage Compressor after overhaul. Compliments of Sun Engineering Services, Inc., and Power & Compression Systems Co.

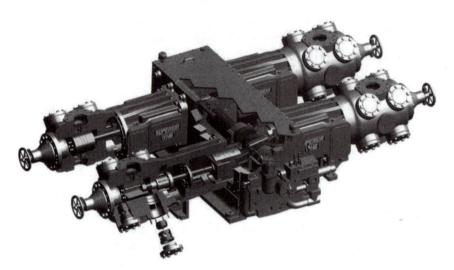

1 Cylinder (head-end), 2 Cylinder (crank-end), 3 Piston, 4 Piston Rings, 5 Piston Rod (Induction hardened in packing area), 6 Compressor Packing, 7 Non-reversible, noninterchangeable valves, 8 Valve Retainer, 9 Valve Cap, 10 Crosshead Guide, 11 Crosshead, 12 Replaceable Tri Metal Shoe, 13 Connection Rod, 14 Main Bearing Cap, 15 Crankshaft, 16 Tie Bar, 17 Breather Cap, 18 Top Cover, 19 Oil Filter, 20 Force Feed Lubricator, 21 Crosshead Guide Support, 22 Suction Nozzle Flange, 23 Discharge Nozzle Flange

Figure 3-42. Superior WH-64 Separable Balanced Opposed Reciprocating Compressor Courtesy of Cooper Cameron Corporation

The separable balanced opposed compressor may be driven by a gas engine, electric motor and, in rare cases, a steam turbine or a gas turbine. Whatever driver is selected a torsional analysis should be performed on the entire drive train. Torsional critical speed and response analysis will be addressed in more detail in Chapter 8, Vibration.

Integral Engine-compressor

The integral engine-compressor is not a new innovation, but in fact, an old one. Although specific dates and manufacturers are not available the integral engine-compressor was derived from the steam engine in the late 1800s. In the mid-1800s a compressor cylinder was fitted to a gas engine. The cross-section of a typical integral gas engine-compressor is shown in Figure 3-43. The concept proved successful for a number of reasons:

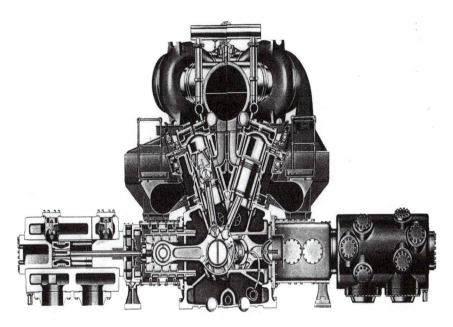

Figure 3-43. Integral Engine-compressor Cross-section with Power Pistons in the Vertical Plane and Compressor Pistons in the Horizontal Plane, Courtesy of Cooper Cameron Corporation

- Manufacturing technology for crankshafts, connecting rods, pistons and bearings was well developed.
- Both engine and compressor shared the same crankshaft and the same lubricating system.
- The need for a coupling was eliminated.
- The base plate was only slightly wider than the engine-only base plate.
- The engine and compressor cylinders could be shipped separately and the compressor cylinders mounted in the field. Thus reducing shipping width.

Preliminary Selection & Sizing

Compressor selection is dependent on target throughput and specific knowledge of the compressor or compressors to be considered. This process starts with determining the specific site and design condition.

Site Conditions

        Altitude = Sea level (barometric pressure)

        Ambient temperature

Design conditions

        Gas suction pressure = P_s

        Gas suction temperature = T_s

        Intercooler temperature = $T_{intercooler}$ = (target 130°F maximum)

        Gas discharge pressure = P_d

        Required flow capacity = MMSCFD (by definition standard
            cubic feet per day is gas flow at 14.7 psia and 60°F).

Gas Properties

        Specific gravity (SpG)

        Ratio of specific heats = k = (see Appendix C—
            physical properties of gases).

In addition to the calculations discussed in Chapter 2, "General Theory," the following calculations are specific to reciprocating compressors.

Determine the number of compression stages by first dividing the final discharge pressure by the initial suction pressure (all pressures must be absolute).

$$\text{Overall compression ratio} = CR = \frac{P_D}{P_s} \tag{3-8}$$

Where

    Pd   = Discharge pressure (psia)

    Ps   = Suction pressure (psia)

The potential stages are determined by taking the root (square root, cubed root, etc) of the CR. The stage CR higher than 4.0 should not be considered as this would exceed material strengths of most compressors. In most cases an acceptable CR will be between 1.3 and 3.9. The higher the CR the fewer stages required and the lower the compressor cost.

Next the adiabatic discharge temperature for each stage is calculated using the following:

$$T_d = T_s \times CR_{stage}{}^{\frac{k-1}{k}} \tag{3-9}$$

Where

T_d = Discharge temperature °R
T_s = Suction temperature °R
CR_{stage} = Compression ratio per stage
k = Specific heat ratio = C_p/C_v.

The maximum allowable gas discharge temperature for general service is 350°F. If calculated temperatures exceed this than use more stages. A good rule of thumb is to select approximately the same CRs per stage.

Note: In determining the number of stages the interstage pressure drop is neglected.

Estimating the required HP is the next step in the sizing & selection process. With this information in hand one or more drivers can be considered. The required horsepower for a compressor depends on the net amount of work done on the gas during the complete compression cycle. The following equation may be used to determine the stage HP or the total HP depending upon the stage information selected.

$$\frac{\text{Theoretical HP}}{\text{MMCFD}} = 46.9 \times \frac{k}{k-1} \times \left[\left(CR^{\frac{k-1}{k}} - 1\right)\right] \times \left[\frac{Z_1 + Z_2}{2 \times Z_1}\right] \tag{3-10}$$

Where

CR = Compression ratio overall or per stage
Z = Compressibility factor overall or per stage

Multiplying the theoretical HP/MMCFD by the throughput required (MMCFD) provides the total Theoretical HP. Figure 3-44 may be used to approximate HP and temperature rise.

Determining the cylinder size, or sizes for multi-stage compressors, is an iterative process that requires knowledge of the cylinder diameter, stroke and speed. These calculations are easily and accurately performed in the manufacturer's computer sizing program. However, the reader can approximate these calculations by utilizing the nomographs in

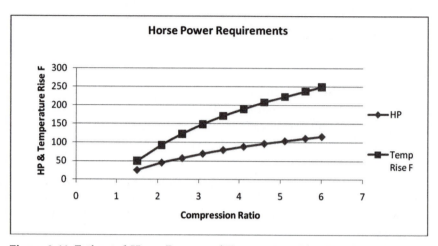

Figure 3-44. Estimated Horse Power and Temperature Rise Based on Compression Ratio

Appendix F—Cylinder Displacement Curves. By selecting the cylinder diameter, stroke and compressor speed the cylinder displacement, in thousand cubic feet per day (MCFD), is easily determined.

With this information in hand the reader can approach any compressor and driver manufacturer and determine if they have the type & size of units required.

Clearance Volume: the volume remaining in the compressor cylinder at the end of the discharge stroke. This volume includes the space between the end of the piston and the cylinder head, the volume in the valve ports, the suction valve guards and the discharge valve seats. This volume is expressed as a percent of piston displacement.

$$\text{Percent Clearance } (\%Cl) = \frac{\text{Clearance Volume}}{\text{Piston Displacement}} \times 100 \qquad (3\text{-}11)$$

Care should be taken that the units used are consistent.

The effect that clearance volume has on volumetric efficiency (Ev) depends on the compression ratio and the characteristics of the gas (that is the "k" value of the gas).

$$\%E_v = 100 - CR - \left[\frac{Z_1}{Z_2} \times CR^{\frac{1}{k}} - 1 \right] \times [\%Cl]$$

$$(3\text{-}12)$$

SCREW COMPRESSORS

One of the earliest, if not the earliest, screw compressors was developed in Germany in 1878 by Heinreich Krigar. The original Krigar rotor configuration resembles the Roots Blower rotor design, which was exhibited in Europe in 1867, with the exception that the Krigar rotors twist through an angle of 180 degrees along the rotor length.

In 1935, Alf Lysholm of Sweden improved the screw compressor with asymmetric 5 female—4 male lobe rotor profile that is still in use today.

The screw compressor is classified as a positive displacement device because a volume of gas becomes trapped in an enclosed space and than that volume is reduced. Within the compressor casing there are two screws with mating profiles, screw "A" and screw "B," with "A" having concave inlets and "B" having convex inlets. These screws rotate in opposite directions with screw "A" receiving power from the outside source and transmitting this power to screw "B" through a set of synchronization gears.

As the screws rotate the process gas is drawn into the inlet or suction port. Gas is compressed by rotary motion of the two intermeshing screws. The gas travels around the outside of the screws starting at the

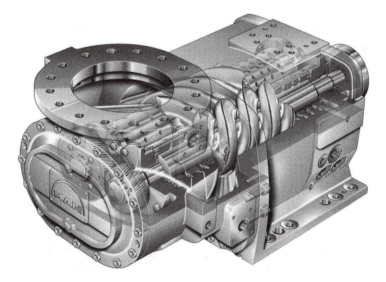

Figure 3-45. Gas Screw Compressor courtesy of MAN Turbo AG

top and traveling to the bottom while it is being moved axially from the suction port to the discharge port. The location of the discharge port determines when compression is complete. A slide valve over the discharge ports is used to control or vary the discharge pressure.

The effectiveness of this arrangement is dependent on close fitting clearances between the two screws and sealing the suction and discharge ports. There is no contact between the screws. To improve compressor efficiency oil is injected into the inlet cavity to aid sealing and to provide cooling. Upon discharge the oil is separated from the gas stream to be recycled after filtering and cooling (as necessary). Separation of the oil from the gas can be achieved down to a level of 0.1 parts per million by weight (ppm wt) with a mesh pad and coalescing elements. Borocilicate microfiber material is typically used in coalescing elements. Depending on the application the gas may be further cooled and scrubbed to remove all traces of oil.

The oil-flooded compressor is capable of flowing over 90,000 standard cubic feet per hour (SCFH) at 200 psig. Injected oil quantities are approximately 10-20 gal/min per 100 HP.

Where oil contamination must be avoided compression takes place entirely through the action of the screws. These oil-free compressors usually have lower pressure and throughput capability. Higher throughputs can be achieved by employing multiple screw compressors in series. In this way throughputs of 120,000 SCFH at 150 psig can be achieved.

In reciprocating compressors a small amount of gas (clearance vol-

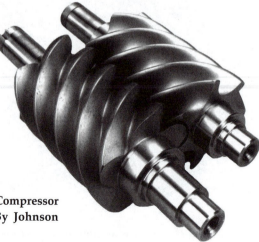

Figure 3-46. Typical Screw Compressor Rotors Courtesy of Frick By Johnson Controls

ume) is left at the end of the stroke. This gas expands on the next suction stroke thereby reducing the amount of suction gas that could have been drawn into the cylinder. At the end of the discharge process of the screw compressor no clearance volume remains as all the compressed gas is pushed out the discharge port. This is significant as it helps the screw compressor achieve much higher compression ratios than reciprocating compressors.

To isolate the screw elements from the bearings, seals are furnished next to the rotor lobe at each end of the machine. Journal bearings are positioned outside the seal area. These are typically hydrodynamic sleeve-type bearings. Tilt-pad thrust bearings are positioned outside of the journal bearings.

Both mineral-based and synthetic oils are used. The oil selected is based on compatibility with the gas being compressed.

Screw Compressor Control

Three of the four control methods used in controlling screw compressors are also used on other compressor types. They are speed control, suction valve throttling and discharge valve throttling. The one control method unique to screw compressors is the slide valve.

Slide Valve

Unloading is achieved by moving the slide valve towards the discharge port (see Figure 3-47). This effectively shortens the working

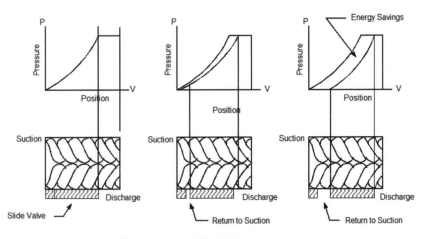

Figure 3-47. Slide Valve Operation

length of the rotor. Since less gas is compressed the required horsepower is reduced.

This control technique is used with a constant speed drive. Flow variation ranges from 100% to 15% with respective power levels down to 40%.

Variable Speed Drive (VSD)

This control technique varies the speed to match the required throughput. As in other rotating element machines attention must be paid to the critical speed of the individual rotating elements.

Suction Valve Throttling

This control technique is utilized in a constant speed application. This technique controls throughput to the desired set point by modulating gas flow into the inlet. While it is the least expensive control method is has it limitations. The two typical control modes are:

- Full Load: The suction valve is open and the unit is making 100% of its rated flow.

- Part Load: the suction valve is partially closed (or open) and restricting gas flow into the suction port.

- No Load: The unit is still turning but the suction valve is closed and there is no flow. As a result the unit is still consuming energy (15-35% of full load) but not producing any flow.

Discharge Valve Throttling

This type of control system is not recommended and is seldom used. If the discharge is closed the compressor will overheat and casing pressure could exceed safe operating limits.

Figure 3-48 summarizes the various control methods vs. power consumed.

Figure 3-49 demonstrates that screw compressors can be very large.

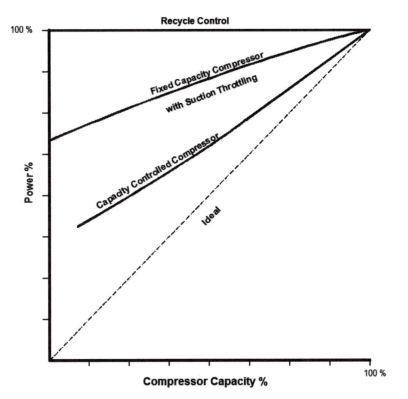

Figure 3-48. Capacity Control vs. Power

Figure 3-49. Screw Compressor Courtesy of *MAN Turbo AG*

Chapter 4

Effect of Operating Conditions

GENERAL

A compressor increases the gas pressure from the point "A" level to the point "B" level in the process of moving the gas from point "A" to point "B." This applies to both dynamic and positive displacement type compressors.

However, the gas is acted upon by its surroundings. Gas composition (molecular weight and specific heat), temperature and pressure influence how a gas behaves. Also effecting compression is the system resistance (the pressure that the compressor must overcome) and the compressor compression ratio (CR). The compressor characteristic curve is a tool in representing how these parameters effect compressor operation. As shown in Figure 4-1 there are different characteristic curves for different pressures and temperatures. This is also true when molecular weight and specific heat are variables. As

EFFECTS OF TEMPERATURE & PRESSURE

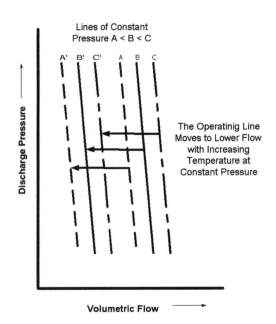

Figure 4-1. Effects of Temperature & Pressure

long as the deviations are small the compressor will be able to adapt to these changes through judicious control techniques.

Considering the effect of gas temperature and pressure from the equation for GHP it is evident that increases in gas inlet temperature, pressure or compression ratio will increase the required GHP necessary to compress the gas. This effect is demonstrated in Figure 4-2.

Also an increase in molecular weight or specific heat ratio will increase the horsepower requirement at constant temperature, pressure and compression ratio as show in Figure 4-3.

DYNAMIC COMPRESSORS

Specific factors that impact the dynamic compressor are surge and the different aspects of surge with varying compression stages.

As design compression ratios are increased the surge line rotates clockwise about the "Zero" point with increasing volumetric flow. (Figures 4-4, 4-5, and 4-6).

Each additional compressor stage changes the shape and location of the surge line thereby generating a composite surge line.

Speed affects axial and centrifugal compressors differently; the characteristic speed curves of the axial compressor being steeper than the characteristic speed curves of the centrifugal compressor. The steeper

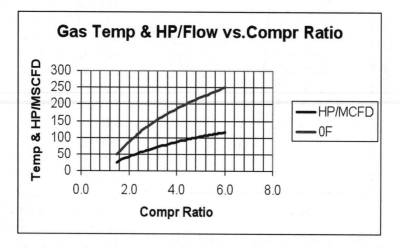

Figure 4-2 Compression Ratio Effects on Gas Temperature & BHP/Flow

EFFECTS OF SPECIFIC HEAT RATIO

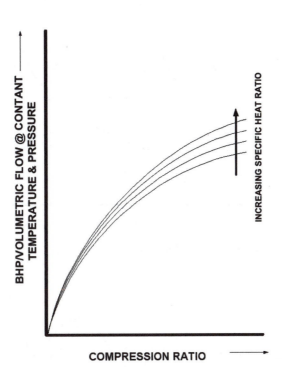

Figure 4-3. Effect of Specific Heat Ratio on BHP/Flow

the characteristic curve the easier it is to control against surge. Whereas the flatter the characteristic curves the easier it is to maintain constant pressure control (or pressure ratio control).

As inlet conditions change the operating point moves along the system resistance curve (assume that the system resistance curve is constant) as follows (see Figure 4-7):

- The operating point moves up along the system resistance curve to Point B as
  — Ambient pressure increases
  — Ambient temperature decreases, and the
  — Surge line moves to the right and closer to the operating point
- The operating point moves down along the system resistance curve to Point A as

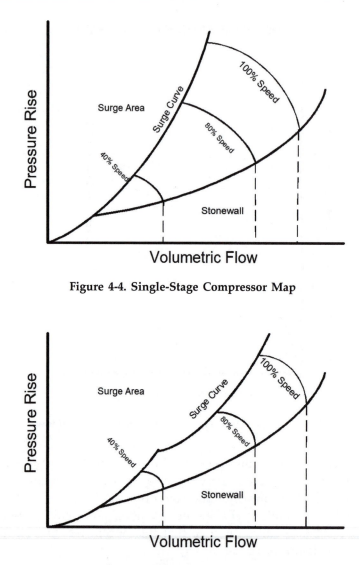

Figure 4-4. Single-Stage Compressor Map

Figure 4-5. Two-Stage Compressor Map

— Ambient pressure decreases
— Ambient temperature increases, and the
— Surge line moves to the left and further away from the operating
     point.

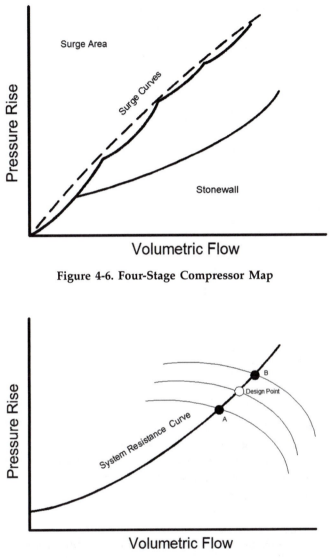

Figure 4-6. Four-Stage Compressor Map

Figure 4-7. Effects of Changing Inlet Conditions

Whereas molecular weight, gas compressibility, inlet pressure and temperature have a significant impact on controlling compressor operation, their impact can be minimized. This is accomplished by recognizing the similarities in the head and flow relationships.

To demonstrate the equations for head and flow can be simplified by considering head versus flow[2]. In so doing the equations for head and flow[2] can be reduced as follows:

$$H_p = \left(\frac{Z_{ave} * T_s}{MW} \right) * \left(\frac{CR^{\sigma} - 1}{\sigma} \right) \tag{4-1}$$

$$Q^2 = \left(\frac{Z_s * T_s}{MW} \right) * \left(\frac{\Delta P_{os}}{P_s} \right) \tag{4-2}$$

since

$$\left(\frac{Z_s * T_s}{MW} \right) \approx \left(\frac{Z_{ave} * T_s}{MW} \right) \tag{4-3}$$

than

$$H_{p'} \approx \left(\frac{CR^{\sigma} - 1}{\sigma} \right) \tag{4-4}$$

and

$$Q^{2'} \approx \left(\frac{\Delta P_{os}}{P_s} \right) \tag{4-5}$$

Where

H_p	$\equiv$	Polytropic head—feet
Z_{ave}	$\equiv$	Compressibility factor—average
Z_s	$\equiv$	Compressibility factor—suction conditions
T_s	$\equiv$	Suction temperature—^0R
MW	$\equiv$	Molecular weight
CR	$\equiv$	Compression ratio

$\sigma \equiv \dfrac{k-1}{k*\eta_p}$—Ratio of specific heats and $\eta_p \equiv$ polytropic efficiency

$\Delta P_{os} \equiv$ Differential pressure across flow orifice in suction line

$P_s \equiv$ Suction pressure—psia

The reduced coordinates define a performance map which does not vary with varying inlet conditions; has one surge limit point for a given rotational speed and compressor geometry and permits or facilitates

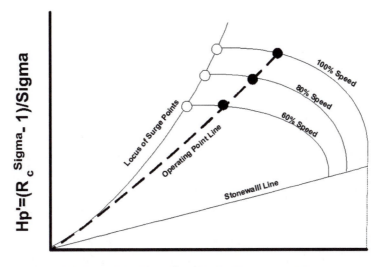

Figure 4-8. Simplified Compressor Map Using Pressures Only

calculation of the operating point without obtaining molecular weight and compressibility measurements.

For a given rotational speed and compressor geometry, the operating point can be defined by a line from the origin to the operating point. This line can be defined by its slope = Hp/Q_s^2. Each coordinate can be reduced by a common factor without changing the slope.

Thus the operating point can be precisely defined by: $(H_p/A)/(Q_s^2/A)$ or $Hp'/Q_s^{2'}$

CENTRIFUGAL COMPRESSOR SAFETY LIMITS

A.   Performance limits (usually overspeed) are not uncommon.
B.   Surge is a serious threat to all dynamic compressors. Virtually every compressor is outfitted with some sort of antisurge protection (bleed valves, VGVs). The consequences are too severe to ignore.
    1.   If the surge limit is crossed, there will be severe high speed oscillations accompanied by vibration and rising gas temperatures
    2.   Surge is a high speed phenomenon—the drop from full flow to

reverse flow can occur in 0.04 seconds.
3. In addition to the surge limit several other performance limits
    (pressure, speed, load, choke) have to be observed as well.
4. Usually choke will not be a problem.

POSITIVE DISPLACEMENT COMPRESSORS

With positive displacement compressors variations in gas composition (molecular weight, specific heat ratio, compressibility), gas
pressure and gas temperature are less critical than they are in dynamic
compressors. While small changes have a minimal effect on compressor
operation, large changes could exceed the available driver power, fail to
meet the required throughput or fail to achieve the required discharge
pressure. In such a case the compressor could continue to operate at
reduced throughput and power. However, failure to achieve the required
discharge pressure would shutdown the operation.

Reciprocating compressors are designed to operate across a range
of pressures or system resistances. As shown in Figure 4-9 the characteristic curve between the high and low system resistance lines is almost
a straight vertical line. At very high differential pressures flow starts to
decrease. Operation in this range usually results in failure (seals, piston
rods, bearings).

One of the major causes of failure in reciprocating compressors is
liquids or hydrates in the gas. Only very small amounts of liquid can
be expelled through the discharge valve. However, prolonged operation
even with a small amount of liquid will lead to valve failure. If the volume of the liquid exceeds the clearance volume either the rod will bend
or the piston will fail. Therefore, it is imperative that the gas entering
the compressor is dry and free from liquids or hydrates.

RECIPROCATING COMPRESSOR SAFETY LIMITS

A. Maximum speed of the driver or the compressor.
B. Performance limits—pressure, temperature, gas composition (mo-
    lecular weight, specific heat ratio) have to be observed as well.
C. Liquids and hydrates in the gas must be avoided

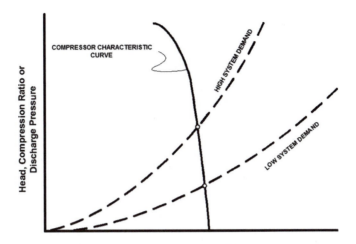

Figure 4-9. Reciprocating Compressor Characteristic Curve

Chapter 5

Throughput Control

Throughput control (capacity control or process control) is achieved by controlling the energy input to the compressor in order to reach the control objective (process set point). For all compressor types this is accomplished by using speed control, suction throttling, discharge throttling or recycle control. In addition to the above, for dynamic compressors, throughput control can be achieved by using guide vane position where applicable. Throughput control for reciprocating compressors can be achieved by using fixed or variable volume pockets and suction valve unloaders. And for screw compressors throughput control is achieved by using a slide valve. A discussion of each of these control methods follows.

SPEED CONTROL

Increasing speed moves the characteristic curve to the right (see Figure 5-1) resulting in increasing flow at a constant pressure or increasing pressure at constant flow. Speed control is the most efficient control method and it can be combined with other control methods for control fine-tuning.

SUCTION VALVE THROTTLING

This control method restricts the gas flow and pressure into the compressor inlet. Suction valve throttling is accomplished by installing a control valve (either linear or equal percentage see Chapter 9) immediately upstream of the compressor and controlling the valve's position as a function of either discharge pressure or flow. As shown in Figure 5-2 suction valve throttling moves the characteristic curve to the left.

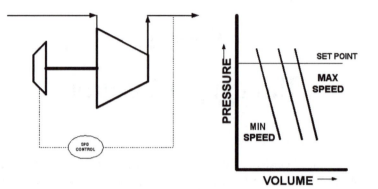

Figure 5-1. Speed Control

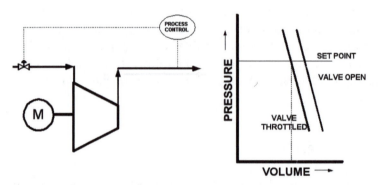

Figure 5-2. Suction Valve Throttling Control

DISCHARGE VALVE THROTTLING

This control method restricts the pressure from the compressor to match the required process pressure at constant flow (Figure 5-3). Because the compressor is working harder than the process requires this control scheme is extremely inefficient. As a result this control technique is rarely used.

VARIABLE GUIDE VANE CONTROL

Variable guide vane (VGV) control applies to dynamic compressor types only. By opening the VGVs the surge point and the characteristic

DISCHARGE THROTTLING CONTROL

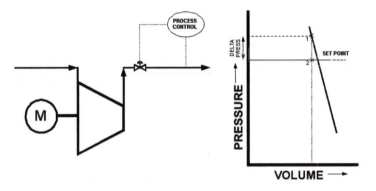

Figure 5-3. Discharge Valve Throttling Control

VGV CONTROL

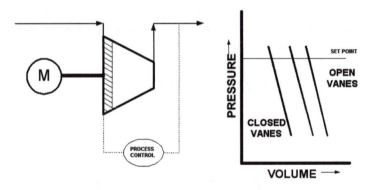

Figure 5-4. Variable Guide Vane Control

curve move to the right. In effect this creates the largest turndown of all the control methods. It also increases the surge margin as the VGV's close. While this control method is more efficient than suction valve throttling it is also more expensive. Not only is the first cost higher for a compressor with VGVs, but also the maintenance cost is higher.

RECYCLE VALVE CONTROL

The recycle valve returns compressor discharge flow back to suction (Figure 5-5). To optimize recycle valve operation a gas cooler is usually installed in the recycle line. In this way the hot discharge gas does not increase the temperature of the suction gas into the compressor. The recycle valve may be modulated from full open to full closed position to provide a 100% range of control. Also a recycle valve may be used with suction valve unloaders to smooth the loading/unloading steps.

SUCTION VALVE UNLOADERS

Suction valve unloaders apply to reciprocating compressors only. Suction valve unloaders prevent gas pressure from building up in each cylinder end. It accomplishes this by holding the valve plates open (finger type unloader) or retracting a plug (plug type unloader) from the center of the valve (suction valve unloaders are discussed in detail in Chapter 3). Suction valve unloaders can be activated to unload one cylinder end at a time or combinations of head-ends and crank-ends as a function of the total throughput required and gas temperature. On some compressor models the gas cycling in and out of the unloaded

RECYCLE CONTROL

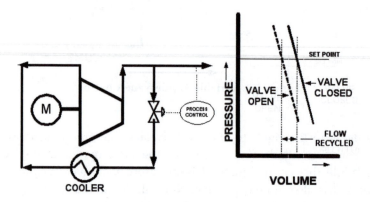

Figure 5-5. Recycle Valve Control

cylinder end can overheat. To address this issue a valve sequence must be created to switch unloaded ends in order to maintain acceptable gas temperatures. Considering a four cylinder single stage balanced opposed separable compressor, each of the eight ends (four head-ends and four crank ends) can be sequenced as demonstrated in Chart 5-1. Ends may be switched, not only to achieve the required throughput, but also to maintain gas temperature.

For example, the four-cylinder single-stage balanced opposed separable compressor in Figure 5-6 is used as a fuel gas booster compressor in parallel with a centrifugal compressor to supply fuel gas to a gas turbine at constant pressure and varying throughput. To meet the gas turbine flow requirements suction valve unloaders where installed on all head-end and crank-end suction valves. These valves are sequenced as shown in Chart 5-1. To achieve smooth flow and pressure transition between each load/unload step the dedicated control program also modulated the recycle valve with each step.

Figure 5-7 shows the interaction of flow, horsepower and suction pressure at each unloading step with changes in suction pressure. At 30 MMSCFD and 190 psig the compressor can operate with 4 ends unloaded and the recycle valve slightly open.

Figure 5-6. Four-cylinder Single-stage Balanced Opposed Separable Compressor

Chart 5-1. Compressor Loader/Unloader Sequence

No Ends Loaded	Sequence within Each Step	Specific Ends To Be Loaded							
		1H	2H	3H	4H	1C	2C	3C	4C
0	N/A	0	0	0	0	0	0	0	0
1	a	X	0	0	0	0	0	0	0
	b	0	X	0	0	0	0	0	0
	c	0	0	X	0	0	0	0	0
	d	0	0	0	X	0	0	0	0
	e	0	0	0	0	X	0	0	0
	f	0	0	0	0	0	X	0	0
	g	0	0	0	0	0	0	X	0
	h	0	0	0	0	0	0	0	X
2	a	X	X	0	0	0	0	0	0
	b	0	0	X	X	0	0	0	0
	c	0	0	0	0	X	X	0	0
	d	0	0	0	0	0	0	X	X
3	a	X	X	X	0	0	0	0	0
	b	0	0	0	X	X	X	0	0
	c	X	0	0	0	0	0	X	X
	d	0	X	X	X	0	0	0	0
	e	0	0	0	0	X	X	X	0
	f	X	X	0	0	0	0	0	X
	g	0	0	X	X	X	0	0	0
	h	0	0	0	0	0	X	X	X
4	a	X	X	X	X	0	0	0	0
	b	0	0	0	0	X	X	X	X
5	a	X	X	X	X	X	0	0	0
	b	X	X	0	0	0	X	X	X
	c	0	0	X	X	X	X	X	0
	d	X	X	X	X	0	0	0	X
	e	X	0	0	0	X	X	X	X
	f	0	X	X	X	X	X	0	0
	g	X	X	X	0	0	0	X	X
	h	0	0	0	X	X	X	X	X
6	a	X	X	X	X	X	X	0	0
	b	X	X	X	X	0	0	X	X
	c	X	X	0	0	X	X	X	X
	d	0	0	X	X	X	X	X	X
7	a	X	X	X	X	X	X	X	0
	b	X	X	X	X	X	X	0	X
	c	X	X	X	X	X	0	X	X
	d	X	X	X	X	0	X	X	X
	e	X	X	X	0	X	X	X	X
	f	X	X	0	X	X	X	X	X
	g	X	0	X	X	X	X	X	X
	h	0	X	X	X	X	X	X	X
8	a	X	X	X	X	X	X	X	X

Each step is to be sequenced as a function of TIME and GAS TEMPERATURE at the compressor end selected.

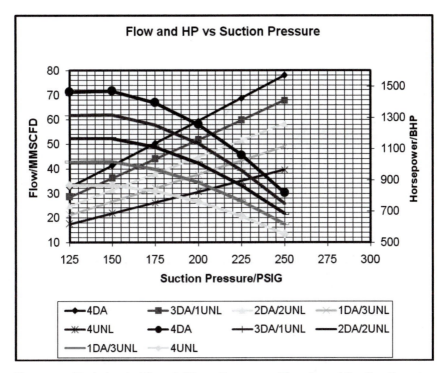

Figure 5-7. Variation in Flow & Horse Power as a Function of Suction Pressure

VARIABLE OR FIXED VOLUME POCKETS

Variable and fixed volume pocket control applies to reciprocating compressors only. Volume pockets vary throughput and efficiency by adding clearance to the piston area. Volume pockets can be either fixed or variable (manual or automatic).

Volume pockets are most easily added to the head-end of the compressor cylinder. While volume pockets could be added to the crank-end the amount of space is often limited. The p-v plot in Figure 5-8 demonstrates how the operation of the volume pockets effects volume and throughput.

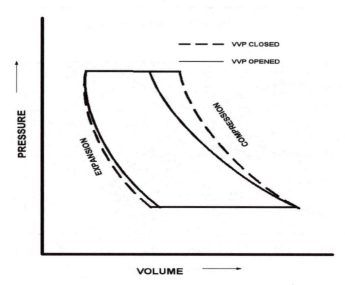

Figure 5-8. Affect of Opening/Closing The Head-End Variable Volume Pocket

Chapter 6

Centrifugal/Axial Compressors Description of Surge

SURGE & STALL

Surge is an operating instability that can occur in dynamic (axial and centrifugal) compressors and blowers (but not in other positive displacement compressors).

Surge and **stall** are often considered to be one and the same. In fact stall is a precursor to surge. As the compressor operating point approaches the surge line a point is reached where flow starts to become unstable.

The stall phenomena occurs at certain conditions of airflow, pressure ratio, and speed, which result in the individual compressor airfoils going into stall similar to that experienced by an airplane wing at a high

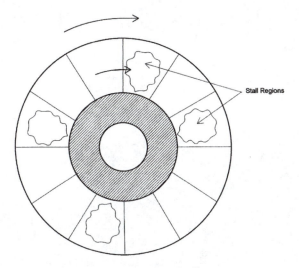

Figure 6-1. Rotating Stall

81

angle of attack. The stall margin is the area between the steady state operating point and the compressor stall line.

As each blade row approaches its stall limit, it does not stall instantly or completely, but rather stalled cells are formed (Figure 6-1). In axial compressors these stall cells can extend from a few blades up to 180° around the annulus of the compressor. Also these cells tend to rotate around the flow annulus at about half the rotor speed while the average flow across each stage remains positive. Operation in this region is relatively short and usually progresses into complete stall. Rotating stall can excite the natural frequency of the blades.

Stall often leads to more severe instability—this is called surge. When a compressor surges flow reversal occurs in as little as 50 milliseconds and the cycle can repeat at a rate of 1/2 to 2 hertz (cycles per second). Although complete flow reversal may not occur in every surge cycle, the fact that flow does reverse has been documented in numerous tests. The most impressive, of which, is when the axial compressor on a jet engine goes into surge (see Figure 6-2). From the point of view of passengers on board the only evidence of surge is the repeated, loud "popping" sound that accompanies surge (unless seated by a window, forward of the engines, and looking directly out at the engine inlet).

In industrial applications flow reversal is observed as fluctuations

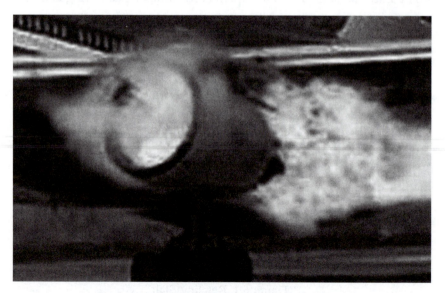

Figure 6-2. Jet Engine Axial Compressor Surge

in measured flow simultaneous with an increase in vibrations. If close to a surging compressor a "whooshing" sound may also be heard.

Surge can cause the rotor to load and unload the active and passive sides of the thrust bearings in rapid succession. When severe the rotor may contact the stators resulting in physical damage.

The compressor map is the single most useful tool for describing surge. The compressor map shows the characteristic curves and the locus of points defining the surge line. The "X" axis is volumetric flow and the "Y" axis may be discharge pressure (Pd), differential pressure (ΔP) or compressor pressure ratio (CR or Rc).

In general, surge can be defined by the symptoms detailed in Table 6-1.

Surge happens so quickly that conventional instruments and human operators may fail to recognize it. This is especially true if the vibration does not trigger the vibration alarm setting. In cases where there was not an "observed incident," the evidence of a surge may be verified by a loss in compressor efficiency.

In industrial compressor applications the severity of surge is de-

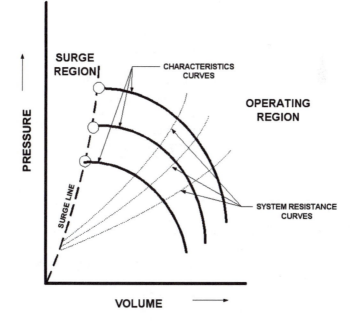

Figure 6-3. Compressor Map

pendent on the volume of the downstream equipment including process vessels, piping, etc. The larger this volume, the greater the damage caused by surge, should it occur. Some factors leading to the unset of surge are listed in Table 6-2.

Table 6-1. Surge Description

1   Flow reverses in 20 to 50 milliseconds
2   Flow reverses at a rate if ½ to 2 Hertz
3   Compressor vibration (mostly axial but vertical & horizontal may also be affected)
4   Gas temperature rises
5   Surge is accompanied with a "popping" or "whooshing sound
6   Trips or shutdowns may occur

Table 6-2. Factors Leading to Surge

FACTORS LEADING TO SURGE

At Full Operation	Trips
	Power Loss
	Jammed Valves (VGVs)
	Process Upsets
	Molecular Weight Changes
	Intercooler Failure
	Rapid Load Changes
	Compressor Fouling
	Dirty Compressor
At Reduced Operation	Start/Stop Cycles
	Load Changes

Some of the consequences of surge are listed below:
- Unstable flow and pressure
- Damage to seals, bearings, impellers, stators and shaft
- Changes in clearance
- Reduction in compressor throughput
- Reduction in efficiency
- Shortened compressor life

Chapter 7

Surge Control
Centrifugal/Axial Compressors

In Chapter 5 throughput control was discussed as it applied to all compressors—dynamic and positive displacement. In this chapter, control will be discussed, as it applies to dynamic compressors only. Specifically this means controlling the dynamic compressor to avoid operating at or near surge. To accomplish this several control techniques will be employed. They are: a) minimum flow control; b) maximum pressure control; c) dual variable control; and d) ratio control.

Before proceeding it is necessary to determine where the compressor is operating relative to the surge region. For a given set of constraints (such as, compressor geometry, available control mechanisms and the process), the compressor operating location can be defined by a line from the origin to the compressor operating point (as was discussed in Chapter 4—Dynamic Compressors).

MINIMUM FLOW CONTROL

Minimum flow control is possibly the simplest and least expensive control method. As indicated in Figure 7-1, the surge margin is normally tuned to prevent surge at high pressures. But when the compressor operates at lower pressure levels, the surge margin is needlessly increased. This is a very inefficient way to control against surge.

MAXIMUM PRESSURE CONTROL

Maximum pressure control is a simple and inexpensive (flow measurement instrumentation is not required) control technique most frequently found in constant speed compressors. This control method is not very efficient and offers only, about, a 10% turndown. As Figure 7-2

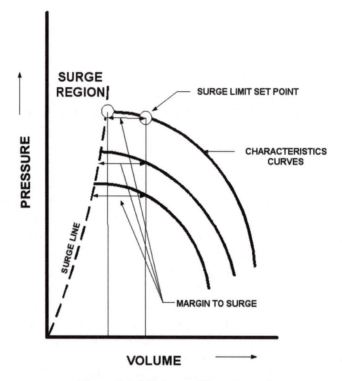

Figure 7-1. Minimum Flow Control

shows, the maximum controller set point does not protect against the operating point crossing into surge at low pressures. And if the operating point goes above the controller set point the recycle valve will open even though the operating point may be far to the right of the actual surge region.

DUAL VARIABLE CONTROL

Dual variable control (Figure 7-3) is employed to control one primary variable while at the same time constraining one or more separate variables. For example, a process may require that the compressor maintain a constant discharge pressure, but at the same time ensure that other parameters (such as maximum motor current, minimum suction pressure or temperature) not be exceeded. Exceeding any of these limits

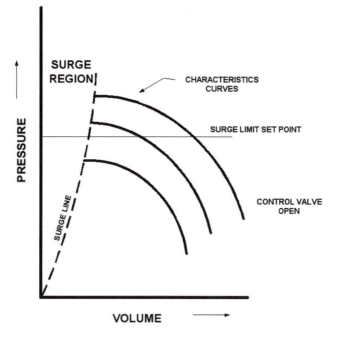

Figure 7-2. Maximum Pressure Control

could result in serious damage to the compressor, damage to the driver
or damage to the process equipment (upstream or downstream).

   Both flow control and pressure control techniques leave the compressor vulnerable to surge as discussed above. A more realistic control
approach would be to establish a control line (or surge limit line) that
mimics the actual surge line and is set up at a reasonable distance (margin) from the actual surge line. This is the approach that has proven most
successful in situations where the above control methods are inadequate.
This method is referred to as ratio control.

Ratio Control
   Recall the equations for polytropic head and flow squared from
Chapter 4 (equation 4-1 & 4-2). The same relationships are rewritten
below in equation 7-1.

$$H_p = \frac{Z_{ave} \bullet T_s}{MW} \bullet \frac{R_c^\sigma - 1}{\sigma} \text{ and } Q = \sqrt{\left\|\left(\frac{Z_s \bullet T_s}{MW}\right) \bullet \left(\frac{\Delta P_{os}}{P_s}\right)\right\|}$$

$$(7\text{-}1)$$

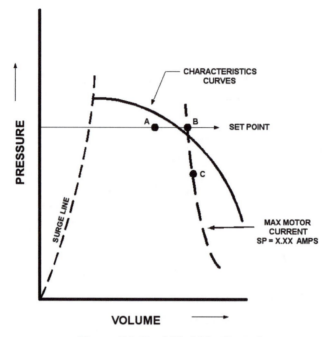

Figure 7-3. Dual Variable Control

Using the format from Figure 7-4, the surge line is renamed the surge limit line (SLL) and the surge limit set point is renamed the surge control line (SCL). To enhance the ability to control against surge over a wide range of conditions additional control ratio limits are added. These are shown in Figure 7-5 Surge Control Map, and listed in Chart 7-1.

The surge control map in Figure 7-5 shows the relationship between the actual surge line, the operating point and various control lines developed to control the compressor throughout its operating range.

These lines are defined as follows:

Using the definitions in Chart 7-1, the operating point can be defined by the slope of the line from the origin to the operating point. This is referred to as the operating point line or OPL.

$$\text{Slope of the operating point line} = \text{OPL} = \frac{H_p}{Q_s^2} = \frac{H_{p1,red}}{q_{s1,red}} \qquad (7\text{-}2)$$

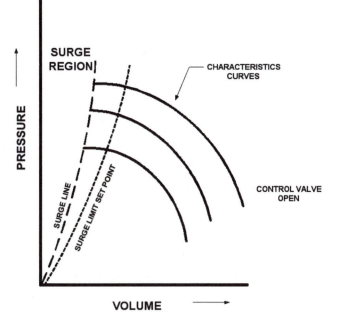

Figure 7-4. Ratio Control

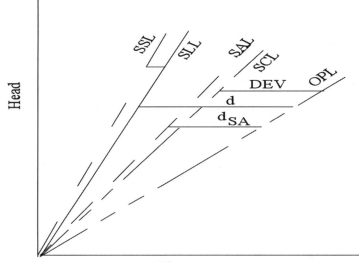

Figure 7-5. Surge Control Map

Chart 7-1. Surge Control Factors

SSL	Slope of the Surge Stop Line
SLL	Slope of the Surge Limit Line
SAL	Slope of the Surge Avoidance Line
SCL	Slope of the Surge Control Line
OPL	Slope of the Operating Point Line @ Design Conditions
b or b_1	Distance Between SCL and SLL
SA	Distance Between SCL and SAL
SS	Distance Between SSL and SLL
d	Distance Between Operating Point and SLL
d_{SA}	Distance Between Operating Point and SAL
T_d	Derivative Action Time Constant
T_{SA}	Time Interval Between Step Outputs
DEV	Distance from Operating Point to SCL

Similarly the surge limit line can be defined by the slope of the surge limit line and is based on the slope of the surge line. The surge limit line is defined as follows:

$$\text{Slope of the surge limit line} = SLL = \frac{H_p}{Q_s^2} = \frac{H_{p1,red}}{q_{s1,red}} \tag{7-3}$$

Furthermore, using the ratio of the two slopes, OPL and SLL a new term is defined which is based of the relationship between these slopes. This term is G_s.

$$\text{Then } G_s = \frac{OPL}{SLL} \ . \tag{7-4}$$

And the distance between the operating point and the surge limit point is defined as "d" where

$$d = 1 - G_s \tag{7-5}$$

When the slopes of the SLL and the OPL are equal, $G_s = 1$ and d = 0. which is as it should be since the distance between the lines is 0.

Also Deviation is defined as the distance from the operating point to the surge control line (SCL). Therefore,

$$DEV = d - b \cdot f(\Delta P_{o,s}) \tag{7-6}$$

where $f(\Delta P_{o,s})$ = function characterizing the shape of the SCL and b is the margin of safety.

$$f(\Delta P_{o,s}) \text{ may be either } \frac{1}{\Delta P_{o,s}} \text{ or } 1. \tag{7-7}$$

If $f(\Delta P_{o,s}) = \dfrac{1}{\Delta P_{o,s}}$ then the SCL is parallel to the SLL and b is the distance between them.

If $f(\Delta P_{o,s}) = 1$ then SCL & SLL intercept at 0 and

$$b = 1 - \frac{SCL}{SLL} \tag{7-8}$$

Using the 0 intercept approach for all control lines the distance between surge control line (SCL) and the surge avoidance line (SAL) is represented by

$$SA \text{ output} = C_1 \left(C_o \cdot d_{SA} + T_d \cdot \frac{G_s}{dt} \right) \tag{7-9}$$

and $b = b_1 + nb_2$ where n is the number of surge cycles.

The above control techniques can now be used to implement algorithms that will avoid a surge and stop surge, should it occur. The surge avoidance algorithm and the stop surge algorithm are described below.

SURGE AVOIDANCE ALGORITHM

The distance from the operating point line to the surge avoidance line is defined by d_{SA}.

Where $d_{SA} = G_s + b_1 + nb_2 - SA - 1,$ (7-10)

and $SA = \dfrac{SAL - SCL}{SLL}.$ (7-11)

When the operating point (i.e. OPL) is to the right of the SAL, d_{SA} will be <0. Only when d_{SA} goes positive is the step output to the bypass valve implemented. Also the step output is repeated in T_{SA} time intervals until the bypass valve is full open. Do not use the derivative if the signal is noisy.

SURGE STOP ALGORITHM

SS is the surge stop line distance and is defined as

$$SS = \frac{OPL}{SLL} - 1$$ (7-12)

When the OPL is to the right of the SSL, SS is < 0 and no action is taken. When the OPL is to the left of the SSL, then SS > 0 and the n value for the number if surge cycles is increased a set amount (n=1, n=2, n=3, etc each time the operating point crosses the surge stop line.

An example in the use of the above surge control technique follows:

CALCULATE THE SURGE LINE SLOPE

Using the following relationship:

$$\Delta P_{os} = \frac{[1-(d/D)^4]}{N^2 C^2 d^4 Y^2} \, \rho \times q^2$$ (7-13)

where $d = 6.373$ and $d = Fa\, d_{meas}$ and $D = 10.02$ and $D = FaD_{meas}$, and Fa is the Thermal Expansion Factor.

$$Fa = 1 @ 70°F \qquad (7\text{-}14)$$

Based on a compressor map the surge point is at the following conditions:

q = 1220 acfm, flow in **actual** cubic feet per **minute**

Discharge Pressure (Pd) = 338 psia

Suction Pressure (Ps) = 118.7 psia

Also gas parameters are as follows:

Molecular Weight (MW) = 17.811

Specific Heat Ratio (k) = 1.286

Compressibility Factor (Z) = 0.98

Suction Temperature (Ts) = 70°F = 530°R

Therefore, **density** $(\rho) = \dfrac{MW \bullet P_s}{Z \bullet R_s T_s} = 0.37929 \qquad (7\text{-}15)$

and

Velocity of Approach Factor (E) =

$$(E) = \frac{1}{\sqrt{1-(d/D)^4}}, \ \beta = d/D \equiv \text{the beta ratio.}$$

$$(7\text{-}16)$$

Rearranging and squaring the above term

$$1/E^2 = [1-\beta^4] = 0.83635 \qquad (7\text{-}17)$$

Gas Volumetric Flow Factor (N) $\equiv$ 5.982114 from Table 9.16 *pp* 9.35*
when flow is measured in acfm.

Discharge Coefficient (C) = $C_\infty + b/R_D{}^n$ $\qquad (7\text{-}18)$

and C_∞ is the discharge coefficient at the infinite Reynolds Number
(table 9/1 *pp* 9.141*).

$$C_\infty = 0.5959 + 0.0312\,(\beta)^{2.1} - 0.184\,(\beta)^8 + \frac{0.09\,(\beta)^4}{D\,[1-(\beta)^4]} - \frac{0.0337\,(\beta)^3}{D} = 0.57874$$

$$(7\text{-}19)$$

$b = 91.70(\beta)^{2.5} = 29.58420$ and $n=0.75$
    and from Table 9.20 equation i pp 9.39* $\qquad$ (7-20)

$$R_D = \left[2266.970\frac{\rho}{\mu DN}\right] q \text{ and } \mu = 0.0106 \text{ @ 1 atmosphere and } 70°F$$
                                                                  (7-21)

from Crane Handbook.

$$\text{Specific Gravity (SG)} = \frac{MW_{gas}}{MW_{air}} = \frac{17.811}{28.966} = 0.61489 \qquad (7-22)$$

From GPSA curves reprinted in Appendix C the critical temperature and pressure of the gas is Tc=362.5°F and Pc=675psia

$$Tr = \frac{70 + 460}{362.5} = 1.46207 \qquad (7-23)$$

$$Pr = \frac{118.7}{675} = 0.17585 \qquad (7-24)$$

Then $\mu = 1.0176 \cdot 0.0106 = 0.0108$ $\qquad$ (7-25)

$R_D = 1,620,432.1$ $\qquad$ (7-26)

$C = 0.57939$ $\qquad$ (7-27)

Gas Expansion Factor (Y) =

$$Y = 1 - [0.41 + 0.35(\beta)^4] \cdot \frac{X_i}{k} \text{ from Table 9.26 } pp \text{ 9.48*} \qquad (7-28)$$

$$\text{Where } X_i = \frac{P_1 - P_2}{P_1} \text{ from eq 9.35, and } k = \frac{c_p}{c_v}$$

    = 1.2863 (iterate as necessary) $\qquad$ (7-29)

---

*Values are found in *Flow Measurement Engineering Handbook*, 3rd Edition, by Richard Miller

Assume surge point at 25 inches of water column then $\dfrac{\Delta P_{os}}{P_s}$

$$= 0.919/118.7 = 0.00774 \tag{7-30}$$

$$Y = 0.99722 \tag{7-31}$$

$$\Delta P_{os} = \frac{[1-(d/D)^4]}{N^2 C^2 d^4 Y^2} \rho \times q^2 = 23.95920 \approx 24 \text{ inches}$$

water column at surge point. $\tag{7-32}$

$$\text{and} \quad q = \sqrt{\frac{\Delta P_{os}}{1.61 \cdot 10^{-5}}} \tag{7-33}$$

Calculate the surge limit line (SLL)

$$\text{Where } \sigma = \frac{k-1}{k} \quad \text{and } k = \frac{c_p}{c_v}. \text{ Then } \sigma = 0.0.22258 \tag{7-34}$$

$$\text{SLL} = \frac{\left[\left(\frac{P_d}{P_s}\right)^{\sigma} - 1\right]}{\Delta P_{os}/P_s} = 5.837 \approx 5.8 \tag{7-35}$$

Similarly the
Surge Control Line (SCL)	= 4.4	(7-36)
Surge Avoidance Line (SAL)	= 4.8	(7-37)
Stop Surge Line (SSL)	= 5.9	(7-38)

And "b" the distance between the SLL and the SCL =

$$1 - \frac{\text{SCL}}{\text{SLL}} = 0.24386 \quad 0.24 \tag{7-39}$$

Chapter 8

Vibration

Every piece of equipment, regardless of size or configuration, has a natural or resonant frequency. Compressors are no exception and, in fact, numerous vibration analysis techniques are employed to predict compressor vibration during the design phase and to identify the source of high vibration in the operation phase. If the resonant frequency (critical speed) is below the operating frequency or speed, the unit is considered to have a flexible shaft. If the resonant frequency is above the operating speed, the unit is said to have a stiff shaft (see Figure 8-1). Almost all axial compressors are considered to have a flexible shaft; that is, the

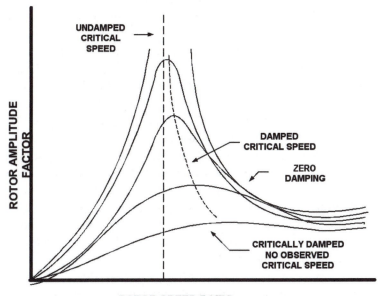

Figure 8-1. Rotor Response Curve Depicting Operation from Critically Damped to Critically Undamped

normal operating speed is above the resonant (or undamped critical speed) frequency. The resonant frequency of centrifugal compressors falls on either side of the undamped critical speed line depending on the number of impellers involved and the shaft size. Blowers, integral-engine-reciprocating compressors and screw compressors generally fall into the stiff shaft category. Separable reciprocating compressors trains (driver & compressor) may fall into either flexible or stiff shaft category depending on the number of cylinders and the type of driver selected. Vibration is a major consideration in the design of compressor rotor assemblies.

Rotor designs vary widely among compressor types and, if fact, they vary even between compressors of similar frame size. Physical factors such as number, size, arrangement of impellers or pistons, and bearing spans, bearing housing design, torque requirements, coupling selection and system parameters such as volume flow, casing pressure rating, operating speed ranges all affect the natural frequency and therefore rotor design.

Each element of the drive system, as well as the complete drive system and all mechanically connected elements, must be subjected to detailed vibration analyses. Regardless of the complexity of the rotor system both lateral and torsional vibration analysis should be addressed. Furthermore, each compressor rotor and support system must be subjected to a thorough dynamic rotor stability analysis. There are a number of excellent companies, large and small, that have the capability to perform the necessary lateral and torsional analysis.

Dynamic rotor stability analysis looks for possible causes of fractional frequency whirl or non-synchronous precession such as a large vibration amplitude component at or near the first lateral bending mode that can lead to excessive rotor vibration. When encountered this vibration can reach magnitudes large enough to move the compressor on its foundation and even destroy the compressor.

Figure 8-2 is a representation of hydrodynamic bearing whirl orbits. Using this technique the rotating shaft journal may be observed as it and the oil wedge moves within the bearing.

The American Petroleum Institute Standard 618 recommends that torsional natural frequencies of the driver-compressor system (including couplings and any gear unit) shall be avoided within 10 percent of any operating shaft speed and within 5 percent of any other multiple of operating shaft speed in the rotating system up to and including the tenth

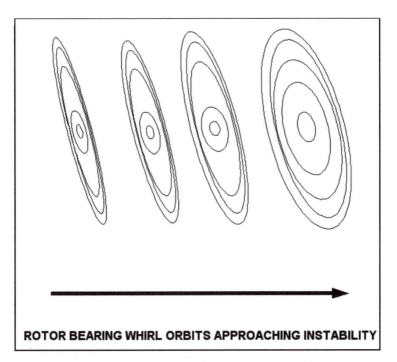

ROTOR BEARING WHIRL ORBITS APPROACHING INSTABILITY

Figure 8-2. Hydrodynamic Bearing Whirl Orbits

multiple. For motor-driven compressors, torsional natural frequencies shall be separated from the first and second multiples of the electrical power frequency by the same separation margins.*

The compressor rotor system (as shown in Figure 8-3 and 8-4), consisting of shaft, impellers, balance drum, etc., together with its support system (that is, bearing type, casing, supports, etc.), is subjected to analysis of its natural lateral frequency response during the design stage. The operating speed range is the most critical element in this analysis. Therefore the speed range should be limited to the predicted operating points without a large safety margin.

Identifying the lateral natural frequencies can be a slow process due to the large number of factors to be considered. Experience is a major asset in this process. For example, a centrifugal barrel compressor usually operates between the first and second lateral bending modes.

*API Standard 618, 4th Edition June 1995 "Reciprocating Compressors for Petroleum, Chemical, and Gas Industry Services," Section 2.5.2

When frequency changes are necessary they can be made by modifying the compressor configuration or stiffening the rotor shaft. Frequency is directly proportional to the square root of the element spring constant and inversely proportional to the square root of its mass as shown in equation 8-1.

$$\omega^2 = \frac{K}{m} \tag{8-1}$$

Where

   ω = natural circular frequency in radians per second,
   k = spring constant or spring modulus or elastic modulus in lb/in, and
   m = mass in lb sec^2/in

Vibration measurements usually consist of amplitude (inches), velocity (inches per second) and acceleration (inches per second per second or "g"). These measurements are taken with probes, such as seismic probes that measure amplitude, proximity probes that measure velocity, and acceleration probes or accelerometers that measure acceleration (see Figure 8-4). By the use of integrating or differentiating circuits, ampli-

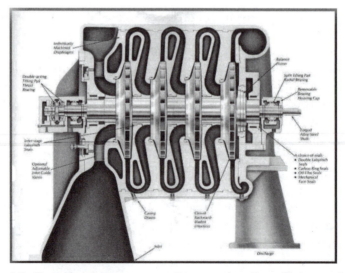

Figure 8-3. Centrifugal Compressor Cross-section Compliments of Dresser Roots, Inc.

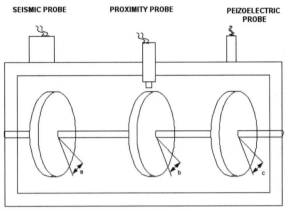

THREE DISK TORSIONAL SYSTEM

Figure 8-4. Simplified Rotor Shaft System

tude, velocity or acceleration readouts are possible from any of the three probe types. However, more accurate information can be obtained from direct measurements using the proximity probe and accelerometer.

Typical sources of vibration that are usually found in a compressor package are detailed in Table 8-1.

Table 8-1. Typical Sources of Vibration

VIBRATION	TYPE	OBSERVATION
0.42 x ROTOR SPEED RADIC	OIL WHIRL	VIOLENT BUT SPO-
0.5 x ROTOR SPEED	BASEPLATE RESONANCE	VERTICAL MORE SENSITIVE
1 x ROTOR SPEED	ROTOR IMBALANCE	
2 x ROTOR SPEED	MISALIGNMENT	AXIAL HIGH
2 x ROTOR SPEED	LOOSENESS	AXIAL LOW
RPM x GEAR TEETH	GEAR NOISE	

Note that rotor imbalance is seen predominantly at the running speed, whereas misalignment can be seen at two times the running speed and is seen on readouts from both the horizontal or vertical probe and the axial probe. Looseness due to excessive clearances in the bearing or bearing support is also seen at two time's running speed, except the axial probe does not show any excursions. Hydrodynamic journal bearing oil whirl is seen at 0.42 running speed and is both violent and sporadic. Very close to that, at 0.5 running speed, is baseplate resonance. Note that baseplate resonance is picked up predominantly on the vertical probe. Gear noise is seen at running speed times the number of gear teeth. Amplitude is not the best criteria to judge acceptable or excessive vibration since a high amplitude for a low rpm machine might be acceptable whereas that same amplitude for a high rpm machine would not be acceptable.

A typical spectrographic vibration plot is shown in Figure 8-5 below. This plot is a "must have" in determining the severity of the vibration and the sources of vibration.

Another useful tool is the vibration map shown in Figure 8-6. This semi-log plot provides RPM and Hertz (cycles per second) on the horizontal scale and vibration in Mils (0.001 inches) displacement, velocity (inches per second) and acceleration (Gs) on the vertical scales. Using this chart it is easy to convert from one system to another. However, for accuracy the following equation is provided to convert Mils to velocity.

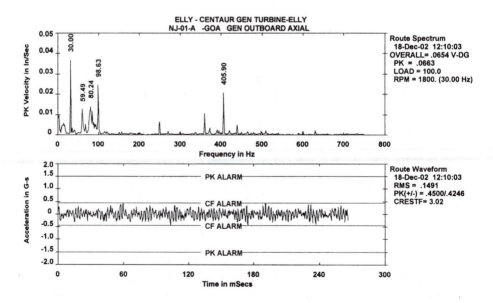

Figure 8-5. Spectrographic Vibration Plot

$$V= \frac{\text{Mils} \times \pi \times \text{rpm}}{60,0000}$$

(8-2)

Where

    V   = Velocity in inches per second

    Π = Pi = Constant = 3.14

  RPM = Revolutions per minute

This chart is also useful in determining when vibration is normal, smooth or rough. As shown on the chart a machine is running smooth when velocity is below 0.03 inches per second (ips), and it is always running rough when velocity is above 0.2 ips.

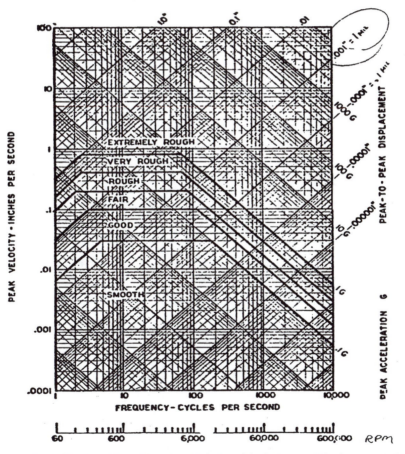

Figure 8-6. Vibration Map Showing Relationship between Displacement, Velocity and Acceleration

Chapter 9

Valve Requirements

Valves have numerous uses in every compressor application. In general these include process control, isolation and safety.

- The process control valve is used to regulate or control the flow, pressure and temperature of the system. The valves that best perform this function are the ball valve, butterfly valve and globe valve.

- The isolation valve is used to "shut in" a compressor. This is necessary in order to perform maintenance on the compressor. The valves that best perform this function are the gate valve, knife valve and plug valve.

- The safety valve is used to protect the compressor and personnel in the area. It is always either automatically activated or is controlled to activate in response to a critical condition. The valves that best perform this function are pressure relief valves and check valves.

Table 9-1 list the primary functions of the most frequently used valves

Valves have different flow-through characteristics as a function of their trim design (see Table 9-2). Valve trims are selected to meet the specific control application needs. In a way this is a throw- back to the days of pneumatic and electric control which were limited in their ability to manipulate valves. While today's computer control capability is sufficiently sophisticated, valve trims are still selected to minimize the programmers work. The most common valve trim characteristics are linear, equal percentage and quick-opening.

Figure 9-1 graphically demonstrates valve trim flow characteristics relative to valve travel. Variations of these three configurations are also available.

Table 9-1. Valve Types

Ball valve	Provides on/off control without pressure drop.
Butterfly valve	Provides flow regulation in large pipe diameters.
Choke valve	Used for high pressure drops found in oil and gas wellheads.
Check valve	A non-return valve allows the fluid to pass in one direction only.
Gate valve	Used for on/off control with low pressure drop.
Globe valve	Provides good regulating flow.
Knife valve	Provides positive on/off control also used for slurries or powders.
Needle valve	Provides accurate flow control.
Piston valve	Used for regulating fluids that carry solids in suspension.
Pinch valve	Used for slurry flow regulation.
Plug valve	Provides on/off control but with some pressure drop.

Table 9-2. Valve Trim

Linear	Flow capacity increases linearly with valve travel.
Equal Percentage	Equal increments of valve travel produce equal percentage changes in the existing C_v.
Quick opening	Large changes in flow are provided for very small changes in lift. Due to its high valve gain it is not used for modulating control but it is ideal for on/off service.

PROCESS CONTROL

Not all processes are the same and different processes require different valve trim and valve response. Throughput control may be accomplished with one or more valves in suction, discharge and/or recycle (throughput control was discussed in detail in Chapter 5). The following recommendations apply to valve selection for all compressor types:

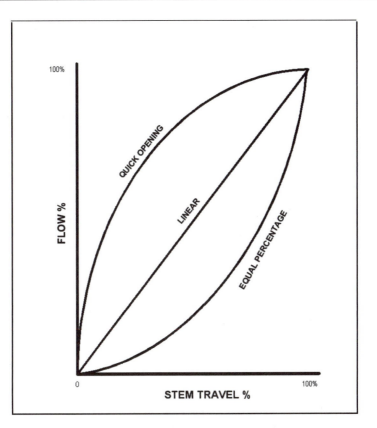

Figure 9-1. Valve Trim Flow Characteristics

- Valve full stroke to be less than 2 seconds.
- Select linear or equal percentage valve characteristics for all but "shut in" applications.
- Minimize the length of pneumatic tubing between the I/P (pneumatic to current) converter and the valve positioner.

For surge control and reciprocating throughput control (for example, using suction valve unloaders) the valve requirements are more stringent:

- Valve full stroke to be 1 second or less.
- Select linear valve characteristics.
- Installation of line booster may be considered as necessary.

Recycle (or blow-off in the case of air compressors) valves are always used with dynamic compressors, but they can also provide flow control in other compressor applications. The following recommendations should be considered for all recycle valve applications:

- Position the recycle valve as close to the discharge port of the compressor as possible.

- Minimize the volume between the compressor discharge, the check valve in the discharge line, and the recycle valve.

- Position the check valve downstream of the "tee" to the recycle valve. The check valve should not be between the compressor discharge and the recycle valve.

- If recycle gas cooling is required, a gas cooler should be positioned downstream of the recycle valve between the recycle valve and the suction line.

- A check valve is an important component of the system to assure safety of the compressor under difficult conditions.

Valves are typically arranged around a compressor as depicted in Figure 9-2. Although both suction and discharge valves are shown, only one valve would be needed.

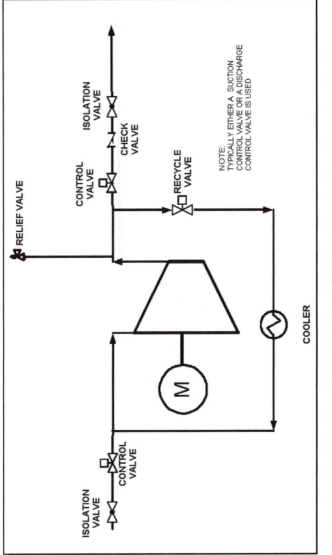

Figure 9-2. Typical Valve Arrangement

Chapter 10

Instrument Requirements

Instruments are devices that measure or control variables such as flow, temperature, liquid level, or pressure. Instruments include:

- Sensors—to measure
  - flow
  - temperature
  - pressure
  - speed
  - vibration
  - liquid level

- Converters to transform the signal from
  - digital to analog or
  - analog to digital,

- Transmitters (to forward the measured signal to local and remote locations),

- Analyzers (combines multiple signals to provide a corrected or related output).

According to the Instrumentation and Systems Automation Society (ISA), formerly known as Instrument Society of America, the official definition of Instrumentation is "a collection of Instruments and their application for the purpose of Observation, Measurement and Control."*

Table 10-A lists some of the most commonly used sensors.

At times more than on sensor is needed to produce an appropriate measurement. For example, an accurate flow measurement requires the

---

*Reference: ISA std. S 51.1—(Instrument Society of America)

Table 10-1. Sensor Types

Parameter	Sensor
Flow	Orifice plate, annubar, rotometer, turbine meter, venturi meter
Temperature	Resistance temperature detectors (RTDs) and thermocouples
Pressure	Manometers, Pitot tubes, piezometers and transducers
Speed	Proximitor probes or magnetic pickups. Proximitors are less likely to be damaged by vibration. Proximitor probes are designed to be intrinsically safe for use in electrically hazardous areas.
Vibration	Seismic, proximitor, accelerometer
Level	Floats

differential pressure across an orifice, the temperature of the gas and the composition of the gas.

The accuracy of flow measurement is especially important in applications such as gas sales. In other applications, such as in dynamic compressor anti-surge control systems and positive displacement throughput control, the main criteria for flow measurement are repeatability and sufficient signal-to-noise ratio. The preferred location of a flow measuring device is in the compressor suction line as close to the inlet port as possible. A less preferable location is in the compressor discharge line as close as possible to the compressor discharge port.

Selection of the location should be based on the following factors:

- Simplification of the algorithm for throughput or surge control
- Simplification of surge protection for dynamic compressors
- Installed cost of the flow measuring device
- Cost of operation of the system as a function of the selected device.

Speed of response is also an important consideration.

- Delay in the control system due to the flow measuring device must be absolutely minimal (head meters or gas velocity).

The flow measuring device and its transmitter should be sized based on the maximum flow of the compressor.

- Select the type and location of the flow measuring device such that the pressure differential corresponding to the maximum flow of the compressor would be 10" water column or more.
- Minimize the length of the tubing between the flow measuring device and the transmitter.

Select the transmitter type and brand based on the following:

- Reliability
- Speed of response (suggest 80 milliseconds or less)
- Thermocouples (in thermowells) are too slow for control or surge detection
- Thermocouples (exposed junction) are preferred for performance trending and surge control
- Resistance temperature detectors (RTD) are accurate in low temperature applications <500°F

Thermocouples and RTDs are made of different materials suitable for different temperature ranges. Thermocouples are classified as type J, K, T, E, R, and S. RTDs are either 10 Ω copper or 120 Ω nickel RTD. Table 10-2 lists the various temperature ranges for each type.

Pressure measurements may be in inches of water ("WC or "H$_2$O), inches of mercury ("Hg), pounds per square in gage (psig), or pounds per square inch absolute (psia). Transducers are the most common instrument used in pressure measurement. Both standard and smart transducers are in common use today. And both have their advantages and disadvantages. Standard transducers provide near-instantaneous readings which are critical for surge control. Smart transducers average the pressure measurements usually over a 1/4 of a second (250 milliseconds). Smart transducers should not be used for surge control.

Table 10-2. Thermocouple & RTD Temperature Ranges

TYPE	RANGE
J	−328° to 1382°F, −200° - 750°C
K	−328° to 2498°F, −200° to 1370°C
T	−330° to 760°F, −200° to 404°C
E	−328° to 1832°F, −200° to 1000°C
R	32° to 3213°F, 0° to 1767°C
S	−40° to 3214°F, −40° to 1768°C
100Ω Pt RTD	−328.0° to 1382.0°F, −200.0° to 750.0°C
10Ω Cu RTD	−328° to 500°F, −200° to 260°C
120Ω Ni RTD	−112° to 608°F, −80° to 320°C

Chapter 11

Detectable Problems

Compressors, either dynamic or positive displacement, can experience a variety of problems, but there are relatively few diagnostic tools to identify them.

Except in cases where vibration is the issue, a thermodynamic analysis of a compressor is the most straightforward way to determine its health. However, this requires that specific instrumentation be installed on the unit. Thermodynamic analysis algorithms were discussed in detail in Chapter 2 - Compressor Theory.

Vibration was addressed in Chapter 8. In addition to the items discussed there, vibration cause and effect is detailed in Appendix H "Troubleshooting."

Fortunately many problems can be avoided before they become costly. Most of these problems can be avoided or minimized with proper and timely preventive maintenance. Obviously not all problems can be avoided. Some problems are inherent in the design. These fall squarely on the shoulders of the manufacturer.

As a rule of thumb a problem that becomes evident during the first twelve months of operation will be covered under the manufacturer's warranty. As used here "manufacturer" also applies to the "packager." The packager is a "third-party" company that installs the compressor, driver (electric motor, gas engine, gas turbine or steam turbine), various vessels, valves, piping, lubrication system and controls on or adjacent to the main skid. The packager, often referred to as the original equipment manufacturer (OEM), has a pre-arrangement with the main component (compressor, driver) manufacturers to package its equipment. Usually this arrangement includes an exclusivity agreement and price discounts. In almost all cases the major equipment warranty is a "pass through" from the specific equipment manufacturer (compressor, driver, etc.).

There are exceptions but once the unit has operated successfully for approximately twelve months any problem that surfaces is usually not covered under the manufacturer's warranty. When equipment has

Figure 11-1. Separable Balanced Opposed 6—Throw Engine Driven Compressor in Pipeline Service Complete with Off-skid Mounted Control System.

not operated successfully during the first twelve months of operation, it is imperative that the owner/operator and the OEM work together to identify and resolve the problem or problems.

When a problem becomes evident, regardless of the compressor's age or run time, the following basic questions should be addressed as soon as possible:

Is the problem mechanical, electrical or performance?

1.   Mechanical problems are associated with vibration, sound, or leaks (air, gas, water, oil). As such mechanical problems are self-evident. Although the cause may not be obvious the symptoms are.

2.   Electrical problems are associated with high voltage to the electric components (motors, motor starters and switchgear) or low voltage to the instruments and control systems. Diagnosing electrical problems is not as easy as diagnosing mechanical problems. However, once the problem is identified the fix is relatively inexpensive.

3. Performance problems usually result from mechanical deficiencies within the compressor (engine or turbine driver and electrical problems with a motor driver). These problems may or may not be immediately evident during operation, but they will lead to equipment failure. This type of failure is usually costly with regards to replacement parts, labor and downtime. Sometimes downtime is the most expensive result of the failure.

Regardless of the type of failure in many cases catastrophic failure can be avoided by applying the following guidelines:

1. Upon installation run the unit to design speed.
    a. Record the unit's vibration signature at several points throughout the expected operating range.
    b. Perform a gas path analysis at several points throughout the expected operating range.
    c. Record temperatures on the compressor not monitored by control instrumentation (such as suction and discharge valve cover temperatures, cooling water temperatures, etc.)
    d. Make note of the different sounds emanating from the compressor during startup, shutdown and at normal operating speed.

2. After the unit has run successfully at a stable temperature for at least four hours shut the unit down and inspect the unit with minimal disassembly (do not remove any close tolerance or close fit components).

In axial compressors surge is the single most severe and expensive failure resulting in damage to the bearings and seals and sometimes to the impellers, diaphragms, blades and vanes as shown in Figure 11-2 and 11-3. Signs of surge are vibration, gas flow instability, gas pressure and temperature fluctuation and audible "huffing" or "whooshing" sounds. Surge occurs very rapidly and it can only be avoided if proper control techniques are applied. Surge was discussed in detail in Chapters 6 and 7.

Compressor Fouling
Indications and Corrective Action

A compressor efficiency drop of 2 percent is indicative of compressor fouling. This can be calculated as shown.

Figure 11-2. Axial Compressor Stator Surge Damage

Figure 11-3. Axial Compressor Rotor Surge Damage

$$\eta_c = \frac{\left(\dfrac{CDP}{P_{in}}\right)^{\frac{k-1}{k}} - 1}{\left(\dfrac{CDT}{T_{in}}\right)^{\frac{k-1}{k}} - 1}$$

$$(11\text{-}1)$$

For the field engineer or operator, compressor fouling is best indicated by a 2 percent drop in compressor discharge pressure at constant speed and load or throughput. Another indication of fouling is a 3-5 percent reduction in load or throughput capacity at constant compressor inlet temperature or ambient air temperature.

Corrective Action
- Dynamic compressor equipment manufacturers normally specify cleaning agents. Typically, a liquid-wash is specified. In extreme cases complete disassembly may be necessary to clean the compressor.

Note: Impact damage to dynamic compressor blades or impellers can mimic compressor fouling. If performance does not recover after washing than the unit should be inspected for internal damage. Many units are equipped with boroscope ports which facilitate internal inspection with minimum downtime and disassembly.

Figure 11-4. Axial Compressor with Contamination Build-up On Blades.

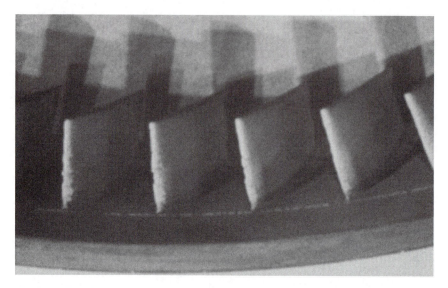

Figure 11-5. Axial Compressor with Contamination Build-up on Leading Edge of Stator Airfoils.

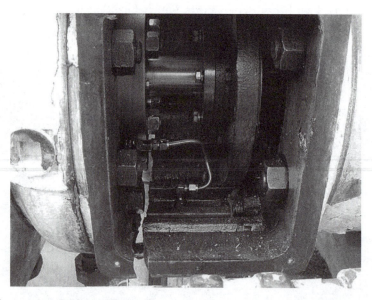

Figure 11-6. Reciprocating Compressor Distance Piece with Piston Rod.

In addition reciprocating compressor fouling (due to either debris or liquids in the gas) may result in damage to the suction or discharge valves and the piston rod. Damage to the valves will manifest itself as an increase in temperature of the valve covers. Rod damage may be detected during operation as non-linear movement of the rod as it passes through the gland at the crosshead-cylinder interface. Also the wear pattern will become more severe as the rod wears against the packing (see Figure 11-7).

Corrective Action
- Disassemble valves, clean and replace valve plates as necessary
- Remove piston and rod assembly and replace rod.

Loose Piston
The installation of the piston rod onto the piston is a delicate procedure requiring a special lubricant, special technique and often special tools. If not properly installed the piston may become loose on the piston rod. This will result in the piston hitting the end wall at the head-end and crank-end of its travel and eventually will result in failure of the piston end wall.

Figure 11-7. Reciprocating Compressor Piston Rod Showing Wear at Location Where Rod Passes through Packing Area.

Corrective Action
- Pull the head-end cover and head-end and crank-end valves to inspect the piston
- If no damage is evident, re-torque the piston rod nut, re-assemble the compressor and test.
- Replace damaged components as necessary

Bearing Clearance (Hydrodynamic Bearings)

Excess bearing clearance can result from improper bearing installation or from journal and bearing shoe wear over a period of time. In either case the effect of bearing looseness can be detected during operation by checking the bearing whirl orbit. Bearing whirl orbits were discussed in Chapter 8, Vibration.

Hydrodynamic bearing whirl orbits should be checked at the factory as part of the unit run test (where the necessary vibration equipment is most likely to be available). At the factory this problem is relatively easy to fix.

Referring to Figure 11-10, the orbit on the left is a stable orbit while the orbit on the right is an unstable orbit. The unstable orbit will also demonstrate high vibration. However, the vibration may or may not be above maximum allowable limits. Continued running with an unstable

Figure 11-8, Compressor Piston on Inspection Table.

Figure 11-9. Typical Hydrodynamic Tilt-pad Bearing. This Bearing Pivots on Pins Specifically Sized to Obtain the Proper Bearing-to-journal Clearance.

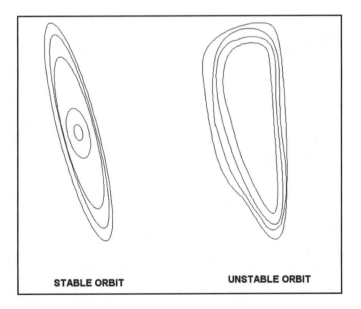

Figure 11-10. Stable And Unstable Hydrodynamic Bearing Whirl Orbits

orbit will lead to increases in vibration that will eventually exceed acceptable limits and, therefore, should not be ignored.

Reciprocating Compressor Valves

Due to the stresses on the valve plates during normal operation suction and discharge valves are the high maintenance items in reciprocating compressors. Valve plate failure will have a negative impact on compressor performance. Gas path analysis coupled with direct measurement of the valve cover temperature is the easiest way to detect valve plate failure.

Corrective Action
- Periodic gas path analysis to monitor compressor performance is the first step in detecting a problem.
- When degraded performance is detected a check of valve cover temperature can isolate the problem valve or valves.
- Replaced the defective valves as soon as possible.

Entrained Liquids

Liquids in the gas present a major problem for reciprocating compressors. As a minimum they will result in early and frequent failure of

Figure 11-11. Reciprocating Compressor Discharge Valve Showing Carbon Build-up.

the compressor valves and when extreme can result in one or more bent rods. Liquids do not present as big a problem for dynamic compressors, screw compressors or blowers as these compressors are more tolerant to liquids in the gas. In fact the screw compressor is so tolerant of liquids that sealing oil is intentionally injected into the gas stream (as in oil flooded screw compressors) prior to compression and extracted after compression. Extremely large amounts of liquids can result in dynamic compressor surge.

Corrective Action
- The first line of defense against liquids is a scrubber or knock-out vessel. This vessel is simply a large area with a coalescing media installed upstream of the compressor where gas velocity is reduced to allow the liquid to separate from the gas. Scrubbers are also installed between stages in multistage compressors.
- Where liquid carryover is severe a cyclone separator may be employed. A cyclone separator utilizes the gas velocity and internal passages to create vortex separation.
- In addition to a scrubber, line tracing or heating may be employed to raise the gas temperature to vaporize the entrained liquids.

Figure 11-12. Reciprocating Compressor Showing Piston Rod Installed in Crank End of Piston.

Figure 11-13. Reciprocating Compressor Showing Failed Piston Rod at Mating Surface with Crank End of Piston.

Variable Volume Pocket Plug

The variable volume pocket plug is positioned via a screw mechanism. Failure of the screw mechanism (see Figure 11-16) could result in the inability to reposition the plug. Screw failure could also result in the plug being sucked into the cylinder where it can become wedged resulting in catastrophic failure (see Figures 11-14 and 11-15).

Corrective Action

• Whether the screw is turned manually or is automated via an electric or hydraulic motor this mechanism should be inspected periodically and maintained per the manufacturer's recommendations.

Lubrication Oil

Lubrication oil should be checked periodically for lubricity, contaminants and water. For small compressors the unit should be shut down and the oil replaced. For compressors with large volume reservoirs the lube oil can be reconditioned on site while the compressor is operating. Lubrication oil should be stored correctly in closed containers with the containers in a rain protected shelter. Figures 11-17 and 11-18 show near-new and failed bearing shoes. The failure was the result of

Figure 11-14. Reciprocating Compressor VVP Failure at Screw.

Figure 11-15. Reciprocating Compressor Showing Failed Distance Piece.

Figure 11-16. Reciprocating Compressor Showing Damage to the VVP Following Failure of the AdjustWer Screw.

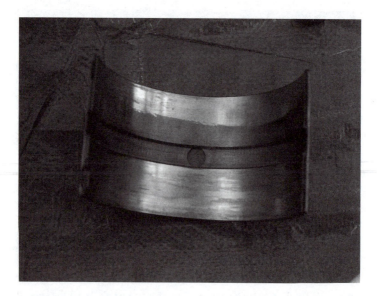

Figure 11-17. Reciprocating Compressor Hydrodynamic Bearing Shoe Showing Normal Wear.

Figure 11-18. Reciprocating Compressor Hydrodynamic Bearing Shoe Showing Distress Following Bearing Failure.

rain water accumulating in the lube oil storage drums and than being introduced into the compressor lube oil reservoir during periodic maintenance.

A NOTE OF CAUTION: The photographs used are taken from actual projects. Therefore, the reader may recognize the manufacturer of the equipment shown in these photographs. However, the problems discussed are not uncommon and may involve any manufacturer's equipment. The reader should not assume that these problems are attributable to a specific manufacturer.

Chapter 12

Controlling Reciprocating and Centrifugal Compressors in Identical Processes

—by Tony Giampaolo

This case study addresses melding together the controllers of two separate compressor booster systems: the dynamic—centrifugal compressor and the positive displacement—reciprocating compressor. This is accomplished through the use of a supervisory control system. The supervisory controller implements the operators' selection for Primary and Secondary compressor systems unless a problem develops. If that happens the supervisory controller can take alternate action based on pre-programmed controller logic.

An edited version of this study entitled "A Unique Approach To Controlling Reciprocating And Centrifugal Compressors In A Fuel Gas Booster System" was published in *Western Energy Magazine* in the summer of 1994.

INTRODUCTION

A unique supervisory control system has enabled Harbor Cogeneration Co., operator of an 80 MW cogeneration plant in the Los Angeles area, to substantially increase the efficiency of its turbine fuel gas booster system. Normally, plant fuel gas is available at or above the 300 psig required by the turbine. However, pressure frequently drops into the 200 psig range, with an occasional excursion as low as 130 psig. During these times a 1,700 horsepower electric motor-driven centrifugal compressor is employed to boost the pressure to at least the required 300 psia.

However, since the centrifugal compressor is designed to operate at a suction pressure of 119 psig, the lowest pressure expected, its efficiency deteriorates as the inlet pressure increases above 119 psig and less compression is required. In other words the centrifugal compressor, which is designed to operate at a compression ratio of 2.3, is actually operating, much of the time, at a compression ratio of 1.1 to 1.3. Centrifugal compressors, when controlled with suction valve throttling and gas recycle, do not significantly reduce the horsepower consumed as suction pressure increases above its design pressure level. In these instances the compressor continues to pump the same amount of gas, mostly into the gas recycle line, resulting in lost energy (approximately 700 BHP) and reduced efficiency. To maintain a satisfactory level of efficiency in the plant fuel gas booster system with varying inlet pressures the company has employed a custom designed supervisory control system to integrate the existing motor-driven centrifugal compressor in parallel with a reciprocating compressor.

The choice of a reciprocating compressor was made after considering a variable frequency drive for the existing centrifugal compressor, or a second, stand-alone compressor.

- The variable frequency drive (VFD) would add another control element—speed control—and would reduce the need to recycle gas at all but the high suction pressure conditions (where recycle would still be required for surge protection). However, the VFD would also add complexity as yet another system: a system without backup, a system subject to failure. In short, the addition of the VFD would improve only efficiency. As a non-redundant system, the VFD would have an adverse affect on reliability or availability. Finally, VFDs are expensive!

- Under varying load conditions, the reciprocating compressor operates very efficiently at constant speed with suction valve unloading. The method provides good flexibility for the price as automatic suction valve unloaders do not add significantly to the package price. In this configuration, and for this application, the reciprocating compressor is slightly more expensive than the VFD. But as a complete and independent system, it is a 100% SPARE! Power savings of up to 690 BHP were expected (and have been achieved) with the reciprocating compressor compared to the centrifugal compressor. However, adding a reciprocating compressor in parallel with the existing centrifugal compressor did present an unusual control interface problem.

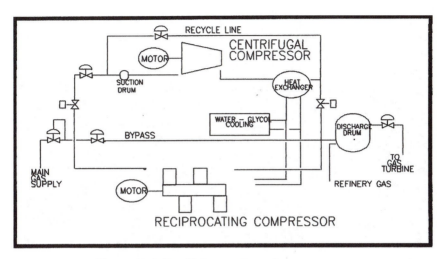

Figure 12-1. Parallel operation of compressors.

DISCUSSION

The reciprocating compressor chosen has automatic suction valve unloading and a small recycle line. The four-cylinder compressor provided eight load steps (two per cylinder), each step representing 12.5 percent of the total throughput. A small recycle line modulates throughput within this 12.5 percent step to maintain smooth flow to the gas turbine. A simplified flow sketch of the centrifugal and reciprocating fuel gas booster compressor in parallel is shown in Figure 12-1.

The 1,500 horsepower motor-driven reciprocating compressor operates at a constant 880 rpm. Throughput control is accomplished with fully automated suction valve unloaders and a modulated recycle line. Both the loaders/unloaders and recycle valve are pneumatically operated via solenoid valves and an I/P converter, respectively. The motor, compressor, frame and cylinder lube systems, cooling system for gas and oil and suction knockout drum are all located on the compressor skid.

THE CONTROL OVERVIEW

In designing this new operating system, Harbor Cogeneration worked with Power & Compression Systems of El Toro, CA. The supervisory control system (Figure 12-4) functions as a central control system. It provides the operator with the ability to pre-select either the

Figure 12-2. Electric motor-driven reciprocating fuel gas booster compressor package

Figure 12-3. Reciprocating compressor control panel

reciprocating or the centrifugal compressor; selection is made at the front of the supervisory control panel (Figure 12-5). Front panel lights indicate which compressor, if any, is running and if either compressor's main breaker has been locked out.

Under normal operating conditions the primary selected compressor is the running compressor. In the event the primary selected compressor is not available (aborted its start sequence, or experienced an unplanned shutdown), the secondary compressor is started. Selection of either compressor as the primary one automatically places the other compressor in the secondary mode.

METHOD OF CONTROL

General

The *start sequence* for the compressors is controlled by the supervisory control located in the main control room. This control was designed specifically by Power & Compression Systems to function as the central control system to integrate the controls of the centrifugal and reciprocating compressors.

It provides the ability to pre-select either the reciprocating or the centrifugal compressor; selection is made at the front of the supervisory control panel (Figure 12-5). Front panel lights indicate which compressor has been selected as *primary*. These panel lights also indicate which compressor, if any, is running and, if either compressor main breaker has been locked out.

Under normal conditions the primary selected compressor is the running compressor. In the event the primary selected compressor is

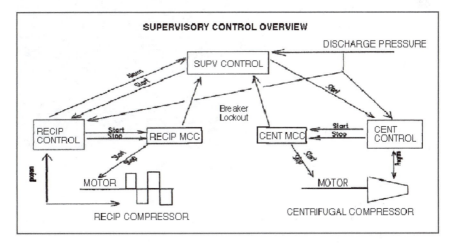

Figure 12-4. Control system interface

not available (aborted its start sequence, or experienced an unplanned shutdown), the *secondary* compressor is automatically started. Selection of either compressor as *primary* automatically places the other compressor in the *secondary* mode.

RECIPROCATING COMPRESSOR CONTROL

The reciprocating compressor is controlled by an Allen-Bradley PLC5-20 mounted in a control panel located next to the compressor skid (Figure 12-3). The PLC controls START, LOAD, UNLOAD, STOP and CAPACITY based on varying suction pressure and constant discharge pressure. The control panel is located off skid to isolate the panel from the vibration common with reciprocating compressors. Using analog and digital signals from the compressor, this control performs data acquisition, monitoring, and diagnostics of the compressor and the control functions.

Figure 12-5. Supervisory control

Automatic Operation

In the "compressor-auto-mode" position the controller initiates the start sequence as a function of a permissive from the supervisory controller, and gas pressure. Once initiated, the start sequence automatically starts the cylinder lube pump and (after lube oil pressures have been established) the main drive motor. The pre-lube pump is set to run continuously to maintain oil temperature and pressure.

The compressor starts with all ends unloaded and the recycle valve in the full open (100%-full recirculation) position. After a warm-up period, the compressor loads one, then two cylinder ends in 30-second intervals until "setpoint" pressure is reached and the recycle valve is modulating between 20 % and 70%. If the compressor remains at this condition for an extended period of time (that is, the recycle valve stays between 20% and 70%), the controller will load other cylinder ends (and unload the initial ends) to avoid heat build-up in the cylinders. This *heat-preclude* sequence is repeated at every load step when the recycle valve movement stabilizes between 20% and 70%. Each load step is initiated when the recycle valve moves to <20% open position. Similarly, each unload step is initiated when the recycle valve moves to >70% open position. At each transition the recycle valve will modulate from the closed to a partially open position to maintain smooth flow to the gas turbine. As suction pressure increases, the cylinder ends will unload in the reverse order.

Manual Operation

In the "compressor-manual-mode" position, the compressor can be started and loaded and unloaded manually. This provision is available for maintenance purposes only.

Remote Terminal Display

Also provided is a remote workstation located in the main control room. This remote workstation interfaces with the field PLC, to provide continuous operating and status information (operating conditions, parameters, and alarms). The remote workstation, a PC-based unit, continuously stores data into memory and makes these data available when requested.

Centrifugal Compressor Control

The centrifugal compressor is controlled by the plant distributed

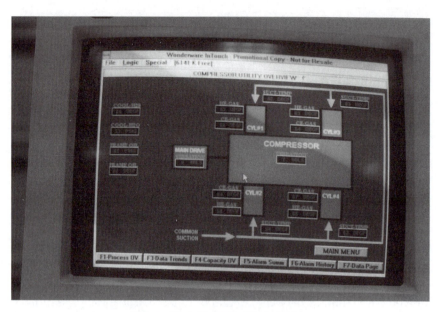

Figure 12-6. Remote Workstation

control system (DCS) and the supervisory control. This control system is augmented by an antisurge controller mounted on the compressor skid.

Automatic Operation

In the "compressor-automatic-mode" position, the compressor initiates a start as a function of gas turbine suction pressure and a permissive from the supervisory control. The controller then modulates the suction throttle valve and the recycle valve to vary gas flow to the gas turbine. The recycle valve also functions as a surge control valve. As suction pressure decreases, the recycle valve modulates to the full closed position. As suction pressure continues to decrease, the suction valve opens.

Manual Operation

In the "compressed-manual-mode" position, the recycle valve must be placed in the full opened position and the suction throttle valve in the near closed (it never goes full closed) position prior to start. Once started, the recycle valve can be closed, and the suction throttle valve opened until the required throughput is reached. Care must be taken that the valve is moved slowly to avoid driving the compressor into surge.

POTENTIAL PROBLEMS THAT WERE AVOIDED

Several potential problems were anticipated and steps were taken in the design stage to address these.

1)   As each cylinder end is loaded, that volume of gas is sent to the gas turbine. This increase in flow could result in over-temperature and/or instability. Either condition would be very detrimental to the gas turbine, and to avoid them the following control elements were utilized:
    a)   Time delays were incorporated into the cylinder loader/unloader logic to facilitate adjusting each sequence during field loop tuning.
    b)   Using open-loop response, the recycle valve was prepositioned prior to each sequence change. Also, time delays were incorporated into the recycle valve logic to facilitate field adjustments.
        During commissioning, these timers were easily adjusted to minimize impact on the gas turbine.

2)   When flow, suction and discharge pressure remain constant, the cylinder ends must be sequentially loaded and unloaded in order to prevent heat build-up in the unloaded cylinders. This "heat preclude sequence" is repeated over the 8-cylinder ends (and throughout the various load steps) so that no cylinder end is completely unloaded long enough for temperature to rise to an unacceptable level. To complicate this task, the torsional study recommended against loading certain end configurations (e.g. 5 ends loaded = with all 4 head ends loaded, load crank end #3 or #4, never #1 or #2).
    a)   Each heat preclude step was programmed into the logic per the results of the compressor manufacturer's torsional study. Since the actual heat build-up time was uncertain (the manufacturer claimed a minimum of 15 minutes per cylinder end) timers were built into each logic step so that they could be easily adjusted in the field. This proved to be more help than originally anticipated, since the heat build-up was actually closer to 10 minutes at the 2 and 3 ends loaded configurations.

3)   The recycle valve had to move fast enough to compensate for the action of the suction valve loader/unloaders. The actuation speed of the loader/unloaders was not known, even to the manufacturer. However, the recycle valve could be made to move, full stroke, in less than one second.

    a) A volume booster was designed into the operation of the recycle valve to improve its response time. Recycle valve "full stroke" speed of response was initially about 5 seconds. This was considered too slow as the loader/unloader responses, when measured, was less than 1 second. To match the responses of the loader/unloader valves and the recycle valve it was easier to improve the response of the recycle valve.

    b) By adjusting the timers in the suction valve loader/unloader logic, valve operation could be staggered slightly.

4)   The supervisory control had to insure that one and only one compressor was allowed to start in automatic. Furthermore, the *secondary* compressor had to start in the event that the *primary* compressor did not start as planned. Finally, regardless of the other compressor's status (shutdown, running, etc.) each compressor had to have the ability to be started manually for maintenance purposes. Figure 12-7 shows some of the ladder logic employed in the supervisory control.

5)   Control loop tuning was accomplished with the aid of a 20-channel high-speed recorder. The recorder indicated the movement of each unloader (16 channels), the recycle valve, and suction and discharge pressure. Initially the control gain and reset rate were set. Once this provided a stable operation at each load level, then the unloaders were also controlled by utilizing open loop control of the recycle steps. The load steps were also controlled by utilizing open loop control of the recycle valve. All of these variations, and adjustments to individual controls, were monitored with the high-speed recorder. The recorder (Astro-Med MT95K2) provided easy programming, calibration, chart speeds selection, and laser printer resolution. The chart speeds utilized were 1 mm/sec, 5 mm/sec, and 25 mm/sec.

    Proportional gain and reset rate was calculated using the Ziegler & Nichols technique for closed loop tuning. Initial and final recycle valve response are shown in Figures 12-8 and 12-9.

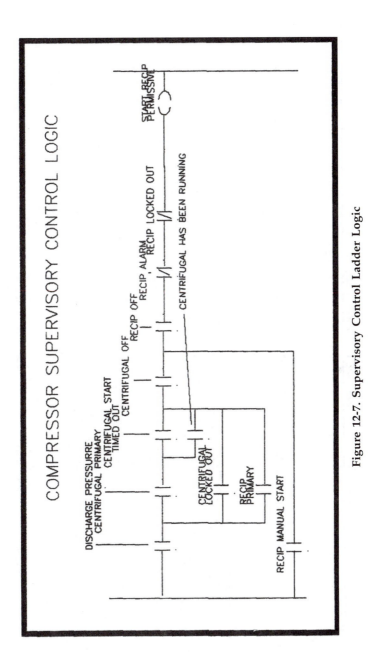

Figure 12-7. Supervisory Control Ladder Logic

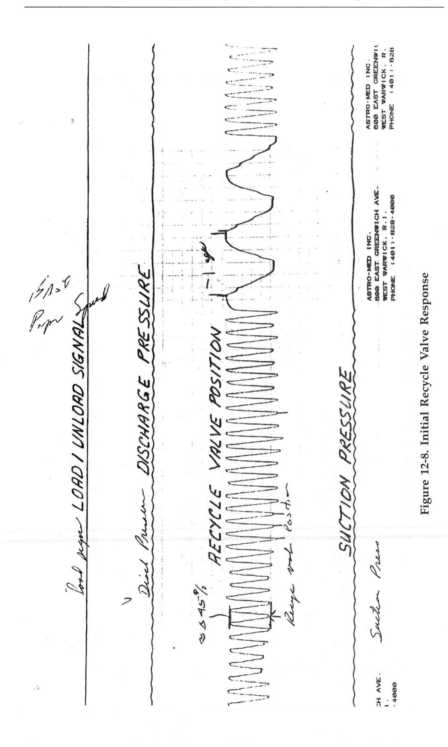

Figure 12-8. Initial Recycle Valve Response

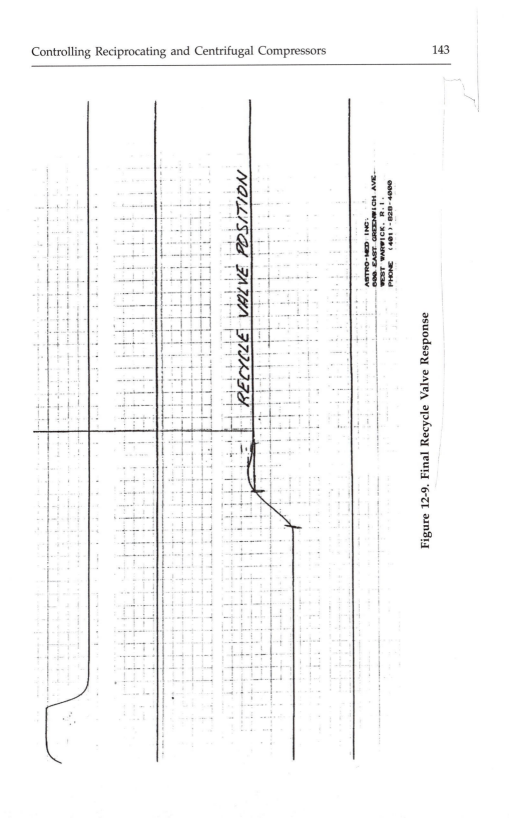

Figure 12-9. Final Recycle Valve Response

CONCLUSION

Performance tests conducted during commissioning confirm the BHP at the design condition was as expected (410 BHP). This is a significant improvement compared to the 1100 BHP required by the centrifugal compressor at the same design conditions.

Variations in pressure ratio and throughput are accomplished by automatic unloaders installed on each suction valve. This reduces the total motor load required compared to the centrifugal compressor at high suction pressures (at low suction pressures efficiencies are comparable). Also each load step (8 steps total) represents 12.5% of the throughput. Therefore, the size of the recycle line and valve are reduced, improving efficiency by eliminating excessive recycle, and the smaller recycle valve contributes to improving valve response time.

While each case is different and should be considered on its own merits, the control techniques employed here can be utilized on compressors (single or multiple configurations, operating in series or parallel) in any environment that requires variation in pressure and throughout.

Chapter 13

Optimization & Revitalization of Existing Reciprocating Compression Assets

—W. Norman Shade, PE
ACI Services Inc.

Compressors, without a doubt, have the longest life expectancy of any piece of mechanical equipment. This study discusses how compressors can be revitalized and returned to service (see also Figures 3-16 and 3-41—Chapter 3—for an example of compressor reconditioning).

Many reciprocating compressors in North America and throughout the world have been in service for decades, some operating for more than half a century. Most older models are no longer optimal, and many are not dependable for current process requirements. Over time, technology has improved and energy costs have increased, making it economically attractive to optimize and revitalize existing compression assets with components that take advantage of modern technology. Efficiency, reliability and maintainability can be greatly improved through custom engineered solutions. Changes may range from conversion to better compressor valves, to the addition of properly designed automatic unloading devices for better control, to the installation of new purpose-built compressor cylinders and components, to the complete reapplication of an entire compression system to a new service.

Even relatively new compressors can become inefficient and mismatched as production needs change. In other cases, from the time the units were installed, actual operating conditions never quite matched the design conditions specified when the equipment was ordered. In still others, manufacturers may have missed the mark when designing and applying their products, so that the equipment underperforms or is unreliable for its intended service.

Whatever the reason, changing valves or cylinders, adding unloaders and automatic control, or reapplying, existing reciprocating compressors is frequently necessary and justifiable. In the case of newer compressors, the OEMs may have standard cylinders that can be retrofitted to their frames, however that usually requires expensive changes to the process piping, pulsation vessels, and cylinder mounting. In the case of mature compressor models that are no longer in production, new OEM cylinders tend to be expensive and have long lead times, if available at all. In some applications, when new compressors are selected or used compressors are redeployed, the standard OEM cylinder line-up may not include the optimal cylinder bore size or working pressure rating for the desired operating conditions. Any of these situations are candidates for custom engineered cylinders and unloading devices that provide cost-effective solutions for preserving the value and reducing the operating cost of existing compression assets.

NEW CUSTOM ENGINEERED REPLACEMENT CYLINDERS

With over 30 years of designing and manufacturing custom engineered compressor cylinders, internal components, valves, and unloading systems, the author's company has helped a large number of compressor users improve their competitive edge by providing creative solutions that improve their existing compression assets to improve performance, efficiency, reliability, maintenance cost, and safety. These efforts have provided solutions to operating problems, have improved the performance and have increased the reliability of reciprocating compressors. The extensive experience in designing and manufacturing custom engineered compressor cylinders is represented in Figures 13-1 and 13-2. Custom engineered cylinders have been applied on all major brands of compressors ranging from small 3.0 in. (76.2 mm) stroke, 1800 rpm high-speed models to giant 20.0 inch (508.0 mm) stroke, 330 rpm frames, with bore sizes ranging from 1.33 to 36.0 in. (33.8 to 914.4 mm), and working pressures from as low as 50 psig (3.5 bar) to as high as 12,000 psig (827.4 bar) for upstream, gas transmission, gas storage, chemical process, refinery, high-pressure air and other services.

New compressor cylinders can be custom engineered and purpose built to meet customer specifications or to solve problems with existing OEM designs. In most cases, the cylinders can be "bolt-in" replacements to match existing process piping connections and mounting locations, which avoids expensive system piping and foundation modifications. Cylinders can be jacketed (water cooled) or non-jacketed (air cooled) and made from appropriate material for the application. Ductile iron castings are most commonly used in recent years, although gray iron, steel or stainless castings are also used, as are steel and stainless steel forgings. Reduced hardness materials are available for sour gas and other applications requiring NACE specifications. Cylinder bores may be supplied in the virgin material condition, ion nitride hardened for improved wear resistance, or lined with a replaceable sleeve. Depending on the requirements, existing internal components may be reusable, however new valves, piston rod packing and maintenance friendly piston and rod assemblies are often included. A wide range of manual and automatically controlled clearance volume pockets and end deactivators are also available, custom engineered for the specific application requirements.

Case History 1: Increased Pipeline Compressor Capacity

In the first example, a mid-continent USA natural gas pipeline operator had several existing Clark HBA and HRA integral engine compressors with cylinders that no longer served the operating needs

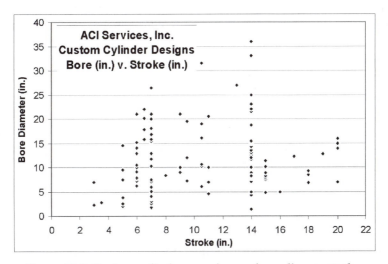

Figure 13-1. Custom cylinder experience—bore dia. vs. stroke

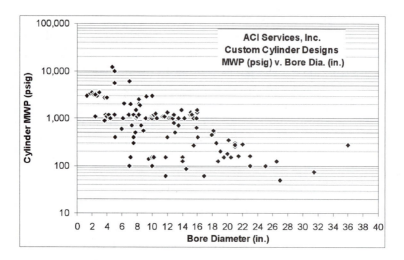

Figure 13-2. Custom cylinder experience—pressure rating vs. bore diameter

of the compressor station. The customer specifications dictated an approximate 32% increase in bore size in a new cylinder that provided maximum operating efficiency. In addition, the cylinder was to have external dimensions that permitted the use of the existing piping system, cylinder supports and foundation. It was readily apparent that a conventional cylinder design would never meet all of the requirements of the specification. The OEM, who supplied the engines, agreed, but offered a conventional design that did not meet the specifications. The Four-Poster™ cylinder offering was sufficiently unique that the author's company received a patent on the new design.

In addition to meeting all the specification requirements, the cylinders were offered at a lower price and shorter delivery than offered by the OEM. In this design, all valves were accessible from the top of the cylinder and the front and rear head, piston assembly, liner and packing case were accessible from the front of the cylinder. A total of 42 new 12.50 inch (317.5 mm) bore, 1000 psig (69.0 bar) cylinders were retrofitted onto multiple 17.0 in. (431.8 mm) stroke compressors. Closed loop tests certified the efficiency of the cylinders. Significant savings resulted from the re-use of the piping and cylinder supports and the absence of foundation changes. Subsequent to the supply of these 42 cylinders, other orders were received for four more identical cylinders and sixteen cylinders with slight modifications for another OEM's 14.0 in. (355.6

Figure 13-3. Retrofitting custom engineered cylinders increased bore size by 32% on this 17 in. (431.8 mm) stroke Clark integral compressor.

mm) stroke engines. All cylinders delivered trouble free service.

Case History 2: Improved Service Life

A USA Gulf Coast chemical plant had a Worthington BDC off gas compressor with cast iron cylinders that were constantly under chemical attack from extremely corrosive gas. Cylinder life was typically only 3 to 5 years, and the OEM had not resolved the problem to the satisfaction of the customer, who required a functionally equivalent cylinder design that would directly replace the existing cylinder to the extent that all other parts of the assembly could be re-used with the new cylinder body.

The customer had attempted to produce replacement cylinders of the same geometry in a ni-resist material, however, the cylinder geometry and jacketed design were not compatible with ni-resist foundry practices and requirements. As shown in Figure 13-4, the author's company redesigned the exterior of the jacketed, 400 psig (27.6 bar), 12.0 in. (304.8 mm) bore, 9.5 in. (241.3 mm) stroke cylinder body to simplify and enhance foundry procedures. Proper core support and clean out and pattern orientation in the mold resulted in sound castings when produced in ASTM A316 series stainless steel.

Figure 13-4. This custom engineered ASTM A316 cast stainless steel cylinder eliminated a severe corrosion problem with the cast iron cylinder that it replaced.

The change to A316 stainless steel resolved the corrosion problems and eliminated the periodic safety inspections that had been required with the original cast iron cylinders. For more than 20 years, the replacement cylinders have been more productive, and maintenance costs have been reduced. Similar replacements of this type were subsequently provided for 19.5 and 7.75 in. bore (495.3 and 196.9 mm) cylinders for adjacent stages.

Case History 3: Providing Safe and Reliable Production Operation

A large Cooper Bessemer V250 integral engine compressor in an Austral-Asian fertilizer production plant had experienced repeat failures of the 3rd stage syn gas cylinder. The 6.50 in. 5 (190.5 mm) lined bore diameter, 20 in. (508.0 mm), 6000 psig (413.7 bar) forged steel cylinder was a design that employed transverse tie-bolts for reinforcement. After several years of service, cracks would originate in the discharge valve pockets and migrate to the tie-bolt clearance holes as shown in Figure 13-5. A thorough engineering analysis revealed very high stresses and unacceptable fatigue safety factors in the valve seats as shown in Figure 13-7.

Leak here

Recycle   Stage 3   Stage 2   Stage 1

Figure 13-5. 3rd stage syn gas cylinder showing location of leak detection.

Since the high-pressure 3rd stage was normally operating at 93% of rated rod load, and the 1st and 2nd stages were lower at 77% and 59%, respectively, a performance study was completed to determine an optimal bore size for the new 3rd stage cylinder. Interacting with the plant engineering staff, it was agreed that the new cylinder bore diameter could be reduced to 6.00 in. (152.4 mm), which cut the normal operating rod load to 84% of rated, while still maintaining acceptable rod loads of 79% and 68%, respectively, on the 1st and 2nd syn gas stages. It was also determined that the low molecular weight permitted a reduction in valve diameter, which significantly reduced the preload on the valve cap bolting, while also increasing the section thickness in the critical corners where the valve pockets intersected the main bore.

A new 6.00 in. (152.4 mm) AISI 4140 forged steel bolt-in replacement cylinder was designed and extensively analyzed to achieve acceptable fatigue safety factors. The new jacketed cylinder, shown in Figure 13-6, was fully machined from a solid block forging that was 3 in. (76 mm) thicker than the original cylinder. The problematic tie-bolts were eliminated, and the critical valve pocket fillet geometry was carefully controlled and then shot peened after machining to provide beneficial compressive preload stresses. Water jacket side covers were fully machined from aluminum plate material to eliminate a corrosion problem that was prevalent with the original cast iron side cover design.

Case History 4: Eliminating Extremely Serious Piston Failures

A major oil company was experiencing frequent and rapid piston failures on the 2nd stage of a reapplied 4000 hp (2983 kW) Superior W76 compressor used for sour gas lift service on an offshore platform. Not only was downtime causing six figure daily production losses, operating personnel were fearful that the piston failures could lead to a breach of the lethal sour gas, which would be extremely dangerous on an offshore

Figure 13-6. New AISI 4140 forged steel cylinder with water jacket, weighing more than 10,000 lb. (4335 kg).

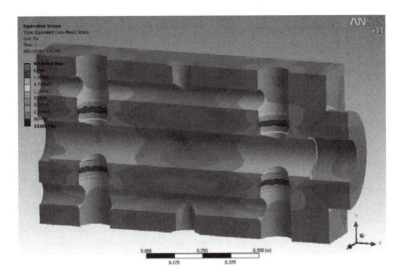

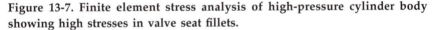

Figure 13-7. Finite element stress analysis of high-pressure cylinder body showing high stresses in valve seat fillets.

platform. The OEM and another compressor service company had both tried to solve the problem with new pistons, but failures continued in as little as 8 days of operating time. Upon being consulted for help, the author's company conducted a thorough assessment and found that very high 2nd stage operating temperature, coupled with rod loads that were at or above the manufacturer's rating, were the root causes of the rapid, catastrophic failures of the 18.0 in. (457.2 mm), 7.0 in. (177.8 mm) stroke multi-piece aluminum and ductile iron pistons. As shown in Figure 13-9, the cylinder was an unusual shape that resulted in a very short piston length relative to the diameter. A finite element stress analysis, as shown in Figure 13-8, showed that the original piston design had a fatigue safety factor of less than 1.0 at the required operating conditions, which explained the rapid failures. Unfortunately, the alternatives for resolving the problem were very limited.

The short cylinder and OEM limits on the permissible reciprocating weight precluded the use of a stronger, less temperature sensitive ferrous material piston. Adding interstage cooling to reduce the 194°F (90°C) suction temperature was not practical on the offshore platform due to space limitations. The OEM did not have an alternative cylinder that would solve the problem, and lead time for a complete new

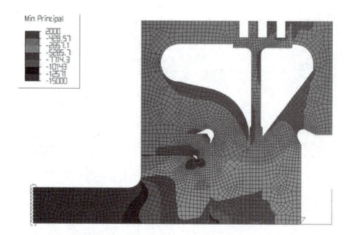

Figure 13-8. Multi-piece Piston finite element stress plot.

Figure 13-9. Original 18.0 in. (457.2 mm) OEM 2nd stage cylinder [left] that experienced frequent rapid, catastrophic piston failures due to a mis-application.

replacement compressor package was about 18 months at the time. A custom engineered replacement cylinder was determined to be the preferred solution, but lead time was also prohibitively long. Concurrent design and analysis efforts established that a one-piece solid aluminum piston, although not having infinite life, would operate for a longer and reasonably predictable period of time. This was adopted as a temporary measure until a new replacement cylinder could be designed and manufactured. With the one-piece pistons providing additional time for development of a permanent solution, further review of the operating conditions of the 1st and 2nd stages showed that further benefits were achievable by changing the two 1st stage cylinders along with the 2nd stage. By changing the two 1st stage cylinders from 20.0 in. (508.0 mm) to 21.0 in. (533.4 mm) and the 2nd stage from 18.0 in. (457.2 mm) to 18.125 in. (460.4 mm), a better balance of rod loads and interstage temperatures was achieved, while providing the compressor with an increased capacity rating that was an unexpected bonus for the operator.

All three of the new ASTM A395 ductile iron cylinders were designed longer to accommodate longer ductile iron pistons that were capable of the required loading. In order to do this while also maintaining process piping connection points, special cast ductile iron offset adapters were designed and manufactured for the suction and discharge connections. The pistons were high-performance, two-piece ductile iron castings with internal ribs. Both the cylinders and the pistons were extensively analyzed with finite element stress models. Standard Ariel variable volume clearance pockets were adapted to the new 1st stage cylinders. The new cylinders received hydrostatic tests at 1.5 times working pressure, followed by helium submersion tests at working pressure. Field-installed on the original frame as shown in Figures 13-10 and 13-11, the new cylinders have completely eliminated the piston failures, while providing the production platform with several percent more capacity than was achievable with the original OEM cylinder configuration.

Case History 5: Increasing Compressor Efficiency and Throughput

A single-stage, gas engine driven, Dresser Rand 6HOS-6 compressor was applied in natural gas pipeline service. The operating conditions of the pipeline were such that the engine driver was fully loaded making it the limiting factor for throughput of the compressor station. Performance tests of the original OEM cylinders, which were misapplied, showed that valve losses were very high for the low ratio application because of insufficient valve sizes and gas passages in the OEM cylinders being used on the frame. New 10.5 in. (266.7 mm) bore, 6.0 in. (152.4 mm) stroke, 1200 psig (82.7 bar) bolt-in replacement cylinder assemblies were designed with larger valves and generous internal gas passages. As shown in Figure 13-12, the new cylinders were installed in place of the original cylinders in less than two days, maintaining all existing process piping connections and mounting. Performance tests of

LEFT: Figure 13-10. New 18.125 in. (460.4 mm) bolt-in 2nd stage replacement cylinder.

BELOW: Figure 13-11. New 21.0 inc. (533.4 mm) bolt-in 1st stage replacement cylinders with manual variable volume clearance pockets.

Figure 13-12. The first of six new high-efficiency bolt-in replacement cylinders [left] is installed on a six-throw, single-stage D-R 6HOS-6 frame. Two of the original OEM cylinders are shown on adjacent throws [right] prior to replacement.

the new cylinders showed an efficiency gain of about 10%, bettering the guarantee of 8% and leading to a corresponding 10% throughput gain for the compressor station.

Case History 6: Increasing Cylinder Working Pressure

A USA natural gas storage facility had a requirement for increased operating pressure. The compression was provided by four Cooper Bessemer GMVA-10 integral engine compressors each having four older style vertically-split valve-in-head cylinders. The OEM, as well as the author's company, was asked to re-rate the original cylinders; however with large gasketed joints as shown in Figure 13-13, the cylinders were only marginally adequate for their original pressure rating and therefore incapable of further uprating. The OEM offered new higher-pressure cylinders, however, they would require major modifications to the ASME coded pulsation bottles and process piping, greatly increasing the cost and the installation time for the project. The author's company measured critical interface dimensions on site without removing the old cylinders and then designed new 12.0 in. (304.8 mm) bore, 14.0 in. (355.6 mm) stroke, 1200 psig (82.7 bar) ASTM A395 cast ductile iron bolt-in replacement cylinders as shown in Figure 13-14. The design was able to accommodate the use of the original double deck valves, packing cases and automatic fixed volume clearance pockets. As the new cylinders were completed, the original valves, packing, and pockets were reconditioned, and the valve pocket volume bottles were re-rated for the higher working pressure. The field conversion went smoothly and the customer's facility was available, on schedule, to handle the higher operating pressure.

Another USA natural gas storage facility required increased operating pressure. The facility had five 1940s vintage Cooper Bessemer GMV compressors with valve-in-head cylinders. As shown in Figure 13-15, the original cylinders had two separate suction and two separate discharge connections, and the cylinders were vertically split so that the outboard suction and discharge flange connections had to be disturbed whenever it was necessary to access the piston, rod or packing for maintenance. Although the OEM indicated that a modest pressure re-rate of the cylinders might be possible, the station operators were concerned about the safety since the existing cylinders sometimes had gasket leakage problems at the original pressure rating. They wanted to avoid any gas piping or pulsation bottle changes and they wanted higher pressure cylinders that eliminated the problematic gaskets and the cumbersome and costly maintenance procedures that required disturbing gas piping connections to access the cylinder bore. In order to meet this challenge, a special valve-in-barrel cylinder design was developed, and 20 new 7.0 in. (177.8 mm) bore, 14.0 in. (355.6 mm) stroke, 2000 psig (137.9 bar) ASTM A395 cast ductile iron, jacketed, cylinders were manufactured and installed on the existing compressors as shown in Figure 13-16. Valves, packing and unloaders were reconditioned and reused in the new cylinders.

Case History 8: Emergency Replacement of Obsolete Cylinders
A chemical plant had an emergency when a critical Pennsylvania ATP off gas compressor unexpectedly developed internal process gas leaks into the cooling water jacket. The OEM no longer had the original cylinder body pattern, so their lead time for a new cylinder approached

Figure 13-13. Original OEM 1100 psig (75.8 bar) valve-in head type cylinders.

Figure 13-14. New custom engineered 1200 psig (82.7 bar) bolt-in replacement valve-in-barrel type cylinders.

Figure 13-15. Original problematic valve-in-head cylinders required disturbing the outboard piping connections to access the piston, piston rod and packing.

Figure 13-16. New bolt-in replacement valve-in-barrel cylinders reused original valves, packing and cylinder unloaders. Piping connections are no longer disturbed for piston, rod, and packing maintenance.

one year. Since loss of this compressor required shutting down the entire plant, emergency temporary repairs were made to the cylinder, and the author's company was contacted to provide a new cylinder. Critical interface measurements were made on site, and a new design 20.5 in. (520.7 mm), 11.0 in. (279.4 mm) stroke, 150 psig (10.3 bar) cylinder was expedited from ASTM A395 cast ductile iron to strengthen the weak, corrosion sensitive sections. The new cylinder was delivered in only 20 weeks, including the time for design, pattern construction, casting, machining, hydrostatic testing and installation of a new ni-resist liner (Figures 13-17 & 13-18).

Case History 9: Extending OEM Standard Cylinder Offerings

An upstream production company had a relatively new Ariel JGJ/6 compressor used for acid gas injection to stimulate oil production. The 600 hp (447 kW), 1500 rpm compressor was originally configured as a 5 stage unit. Once the field was operating, however, it became evident that production could be enhanced by increasing the injection pressure to a level that exceeded the capability of the compressor as originally configured. It was determined that a 6th compressor stage was required, but the OEM did not have a suitable high-pressure cylinder for that

Figure 13-17. Liner installation.   Figure 13-18. Finished cylinder.

particular frame.

In addition, the aggressive nature of the acid gas required the use of material produced to NACE specifications; and the limited physical size constraints required a higher material strength than could be achieved with the use of reduced hardness alloy steels. 17-4PH stainless steel was therefore the material of choice for the new cylinder. With the approval of the OEM, the author's company designed and manufactured a special 2.75 in. (69.9 mm) bore, 3.5 in. (88.9 mm) stroke, 3000 psig (206.8 bar) cylinder. The tail rod design, which is shown in Figure 13-19, was completely machined from a solid 17-4PH double H1150 heat treated stainless steel forging. This cylinder was added to the original compressor as a 6th stage to increase final injection pressure.

Case History 10: Filling gaps in OEM Standard Cylinder Offerings

A distributor packager of new Gardner Denver and remanufactured Joy 7.0 in. (177.8 mm) stroke high-pressure air compressors used in portable diesel engine driven compressor packages for the rotary drilling industry was encountering application conditions for which the OEM cylinders were not optimally suited. Given the maturity of the compressor product line, the OEM was not very interested in modifying the existing cylinder designs or creating new cylinder designs.

The author's company reviewed the range of required application conditions, measured the mounting interface dimensions, and subsequently developed a family of new 7.0 in. (177.8 mm) cylinders that fit the compact vee-type Joy and Gardner Denver compressor frames and that were optimally sized for the air drilling application requirements.

Figure 13-19. Special 3000 psig (206.8 bar) 17-4PH stainless steel tail rod cylinder designed for acid gas injection service.

The bore sizes of the ASTM A395 cast ductile iron, jacketed cylinders, ranged from 3.5 to 4.5 in. (88.9 to 114.3 mm) with a working pressure of 2500 psig (172.4 bar). Cylinders are now supplied on a regular basis to meet this distributor packager's needs, and additional bore sizes are being considered to further expand the range of pressures and flow rates offered to meet the needs of the drilling market as shown in Figure 13-20.

Many more examples can be provided, however, the forgoing list shows the various kinds of applications for which new custom engineered, purpose built reciprocating compressor cylinders can be attractive alternatives for optimizing and revitalizing existing compression assets.

RE-APPLIED COMPRESSOR CYLINDERS

Although new, custom engineered and purpose-built cylinders can provide optimal performance and reliability, there are frequent

Figure 13-20. 2500 psig (172.4 bar) cylinder family developed to fill a gap in the OEM's offerings for high-pressure air drilling service.

examples where surplus or previously decommissioned cylinders can be reconditioned and reapplied onto existing compressor frames. New OEM cylinders can also be installed on different OEM frames that have similar strokes and rod diameters. Of course, diligent and experienced engineering design is required to match the different mounting, bolting, piston rod and, if the stroke is different, the piston. Consideration must also be given to matching reciprocating weights with the compressor frame's capability. Maximum pressure ratings, Internal gas and inertia rod load ratings, and pin reversal requirements must be considered in the reapplication as well. Done properly, reapplication of cylinders is a reliable alternative that can often reduce the capital cost and will almost always reduce the cylinder lead time. Several examples demonstrate the potential of this engineered approach.

Case History 11: Adapting Remanufactured Cylinders to An Existing Compressor

A natural gas pipeline transmission station had two 1950s vintage I-R KVG 10/3 integral engine compressors in service. Although having a provision for three compressor throws, only two of the throws were previously equipped with cylinders, the center throw being unused for more than 50 years. Conditions had changed over that period of time so that the 600 hp (447 kW), 330 rpm compressors were no longer fully loaded. The pipeline operator had determined that, by adding another identical cylinder to the inactive throw in parallel with the existing cylinders in a single-stage configuration, the engine could be fully loaded to increase the station's capacity. Inquiries were made to the OEM and several other companies. The OEM still had the original pattern and was able to quote new identical cylinders, however lead time for that option was about 40 weeks. The author's company also offered to design and manufacture new bolt-in replacement cylinders, however, the best possible lead time was 26 weeks. Neither of these alternatives were adequate for the operator's requirement to have the conversion completed so that both compressors would be in service within no more than 20 weeks after the release of the order. An extensive search for identical used or surplus cylinders produced no results either, although the author's company was able to find two similar 15.0 in. (381 mm) stroke I-R process cylinders, Figure 13-21, that had the same flanges and distance between flanges. It was determined that with modifications to the combination crank end head and distance piece, the process cylinders could be adapted to fit the existing KVG compressors. A performance model, Figure 13-22, showed that the cylinders would meet the required operating conditions if all cylinders were modified by lining to a 10.16 in. (258.1 mm) bore diameter.

Having no time to lose, two used cylinders were purchased and taken to the factory for cleaning, inspection and hydrostatic testing to confirm that the cylinder bodies were sound. The cylinders were then modified to fully restore all fit dimensions, sealing surfaces and threads to ensure their integrity for the reapplication. All new bolting, seals and gaskets were installed. After machining, a second hydrostatic test of the remanufactured cylinders was conducted at 1.5 times maximum working pressure, as shown in Figure 13-23. Concurrent with the completion of the "new" remanufactured cylinders, new liners, non-lube piston and rod assemblies, double deck poppet valves, valve cages, valve caps, outer end automatic fixed volume clearance pocket heads, and highly efficient radial poppet suction valve deactivators were designed and manufactured for the four existing cylinders on the two compressors as well as for the two remanufactured cylinders. A complete API 618 design approach 2 pulsation study was also conducted on the entire system

Figure 13-21. One of two used I-R cylinders prior to the remanufacturing process.

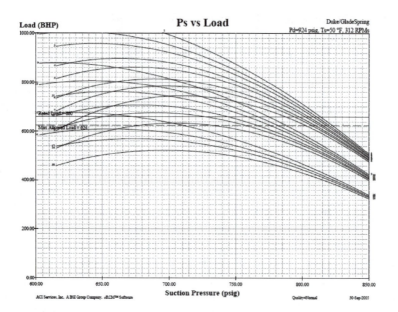

Figure 13-22. Performance model for 3-cylinder, single-stage I-R compressor showing all load steps required for system flexibility without overloading at the maximum specified discharge pressure.

with the new configuration. This study revealed high pressure drops in the existing plant suction pulsation control arrangement with the higher flow, three cylinder configuration, and mitigating actions were quickly identified and added to the conversion plan.

When all components and cylinder assemblies were completed, the station was taken off line for the conversion. The suction and discharge bottles were removed and taken to an ASME code welding shop for addition of the incremental connections, followed by recertification. Some simple piping modifications were also made to mitigate the pressure drop problem that had been revealed by the pulsation study. Using special field machining equipment, the existing liners were machined out of the existing cylinders without removing them from the compressor. New liners and the other new components were then installed along with the incremental remanufactured cylinders. The revised bottles were reinstalled and the system was completed and commissioned on schedule. One of the completed units is shown in Figure 13-24. At this time the compressors have been in service for over 3 years.

Figure 13-23. Remanufactured cylinders with new fixed volume clearance pocket heads and valve caps undergoing an 8-hr factory hydrostatic test.

Figure 13-24. One of two reconfigured KVG units after adding a 10.16 in. (258.1 mm) diameter remanufactured cylinder [center] and adding new clearance pockets, radial poppet suction valve deactivators, and other new components to all the cylinders.

Case History 12: Mixing Used OEM Cylinders and Frames to Optimize Performance

A natural gas processing plant wanted to reapply an existing 2000 hp (1491 kW), 900 rpm gas engine and compressor package for a propane refrigeration process. Process conditions were fixed, so that cylinders had to be carefully selected to match the operating requirements. Using the preferred 6.0 in. (152 mm) stroke Superior WH64 compressor frame, initial performance sizing resulted in a selection of 25.5 in. (648 mm) diameter Superior cylinders for both the 1st and 2nd stages, however the 2nd stage cylinders would have had to operate very close to their working pressure limit. The same concern existed for the 20.0 in. (508 mm) 3rd stage cylinder. The OEM was not interested in producing special cylinders for a single unit, and no used cylinders with an adequate working pressure could be found.

As a result, a new 6.5 in. (165 mm) stroke Ariel JGC/4 compressor was considered, but this resulted in concern about the reliability of the valves with the heavy propane gas. Switching to a 5.5 in. (140 mm) stroke Ariel JGD/4 compressor made the valve reliability more promising, but it could not produce the required flow with the available standard Ariel cylinders. It was finally determined that the solution could be optimized by adapting Ariel JGD cylinders to fit on the Superior WH64 compressor frame. Ariel 22.0 in. (559 mm) diameter JGD cylinders were selected for the 1st stage and the two 2nd stage cylinders, and an Ariel 19.625 in. (499 mm) diameter JGD cylinder was selected for the 3rd stage. Used Ariel cylinders were found that could be reconditioned for all but one of the four that were required, however since the JGD cylinders are designed for a 5.5 in. (140 mm) stroke, their use on the Superior frame required redesigning the piston and piston rod to accommodate the longer stroke. In addition, special mounting adapters were required to match the different bolt patterns on the cylinders and the frame, as shown in Figure 13-25. The author's company provided the special engineered components to accomplish this adaptation. At this time, the unit, shown in Figure 13-26, has been in service for approximately 3 years.

Case History 13: Extending Obsolete Compressor Life with New OEM Cylinders

A natural gas production company had an obsolete 6.0 in. (152.4 mm) stroke Knight KOCA-2 compressor frame with fabricated (welded) steel cylinders, as shown in Figure 13-27. The cylinders were susceptible

Figure 13-25. Adaptation of 5.5 in. (140 mm) stroke Ariel cylinders to a Superior 6.0 in. (152 m) stroke Superior compressor frame.

Figure 13-26. 2000 hp (1491 kW), 900 rpm, gas engine driven 3-stage propane refrigeration compressor.

to cracking problems in service and were no longer safely repairable. The original compressor OEM had been out of business for more than two decades, so neither new nor repairable used cylinders were realistic alternatives. Lead time for new compressor packages was close to 52 weeks and production losses would be substantial if the compressor could not be returned to service in a much shorted time. The author's company was asked to evaluate alternatives. Although new custom engineered cylinders were an option, lead time was still somewhat longer than desirable, considering the cost of lost production. A much shorter lead time was possible by adapting new Ariel JGK cylinders to the KOCA frame. As long as changes were going to have to be made to the cylinders, bottles and piping anyway, the producer decided that it would be desirable to optimize the compression system to accommodate an anticipated suction pressure decline over time, while still utilizing as much of the gas engine driver power as practical to deliver flow. This ultimately required converting from a single-stage to a 2-stage configuration with 8.375 in. (212.7 mm) and 6.25 in. (158.7 mm) bore cylinders, which provided optimal coverage of the operating conditions and maximum driver power utilization as shown in Figure 13-28. Even more importantly in this case, the solution could be implemented in less than 20 weeks time.

New 5.5 in. (139.7 mm) stroke Ariel cylinders were converted to 6.0 in. (152.4 mm) stroke with new custom-engineered piston and rod assemblies, rod packing and wiper assemblies, and mounting adapters to match the KOCA frame. New ASME coded pulsation bottles were also designed and manufactured, as shown in Figure 13-29, and piping modifications were made to adapt to the original skid and utilize the process gas cooler. An API 618 Design Approach 2 pulsation study was conducted, and the system piping design was tuned to provide effective pulsation control over the broad operating range. The unit was been in service for more than two years, and additional units are now being considered for similar conversions.

Case History 14: Completely Reconfiguring an Existing Compressor

An existing 600 hp (447kw), 588 rpm, 4-throw, 3-stage, 7 in. (177.8 mm) stroke Joy WB14 compressor was originally configured as shown in Figure 13-30 to compress 1303 cfm (42.5 m3/min) of bone dry nitrogen. Requirements changed such that only 314 cfm (10.3 m3/min) was required. The cost of recycling 76% of the flow was not acceptable, but

Figure 13-27. Obsolete KOCA-2 compressor prior to conversion.

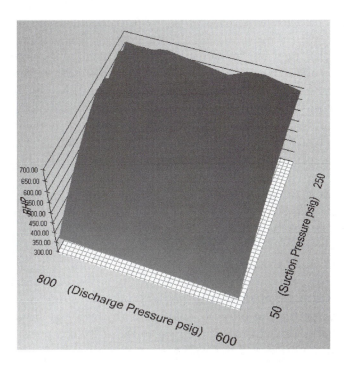

Figure 13-28. Loading map for new 2-stage configuration.

Figure 13-29: Obsolete KOCA/2 frame with new Ariel cylinders applied.

the cost of a complete new, but smaller, machine was equally difficult to justify. The owner/operator wanted to reconfigure the existing compressor at minimal cost to operate more efficiently at the lower flow rate. In order to evaluate possible alternatives, a complete performance model was developed for the existing configuration having two 16.0 in. (406.4 mm), one 10.5 in. (266.7 mm) and one 7.0 in. (177.8 mm) cylinders. The model was then revised in an iterative manner to evolve a solution that let replacing the two 1st stage cylinders with reconditioned used 8.0 in. (203.2 mm) cylinders, moving the original 7.0 in. (177.8 mm) 3rd stage cylinder to the 2nd stage, and installing a reconditioned used 4.0 in. (101.6 mm) cylinder on the 3rd stage. The bottles were modified with new flanges to adapt to the smaller cylinders. Most of the existing piping was reused by adding adapter sections.

By reconfiguring the compressor with reconditioned used cylinders and reapplying one of the existing cylinders, as shown in Figure 13-31, this solution provided a cost effective solution that saved significant operating cost and preserved the original investment of the compressor frame, motor and the balance of the system.

Figure 13-30: Prior to reconfiguring, this oversized 3-stage Joy WB14 nitrogen compressor required 76% of the flow to be recycled.

Figure 13-31: The reconfigured unit was sized to meet the process requirements without significant recycled flow.

COMPRESSOR OPTIMIZATION AND AUTOMATION WITH CUSTOMIZED UNLOADING DEVICES

In addition to hundreds of successful examples of custom designed new cylinders and countless innovative adaptations and reapplications of new and used OEM cylinders, there are even more examples of compressor optimization with automatically actuated devices, often called unloaders, that control the load and flow.

Case History 15: Automation of Propane Refrigeration Compressor

A natural gas processing plant used a 400 hp (298 kW) Ajax compressor with no automatic control. The 2-stage, 2-throw unit served as the low-pressure stage of a multi-unit propane refrigeration system. As ambient conditions changed, the propane boil-off rate varied, necessitating manual adjustments of the compressor's clearance pockets. Since the unit was not attended 24/7, the operators kept the unit under-unloaded so that it could handle upsets without shutting down. Unexpected shutdowns stopped the liquid stripping process and led to serious problems downstream from the gas plant.

The solution involved the combination of a discrete step unloader and an infinitely variable unloader providing, in essence, infinite variability over a broad range of operating conditions, as shown in Figure 13-32. This gave the plant maximum compressor loading (and maximum plant refrigeration and liquid stripping capacity) without the risk of unexpected shutdowns.

The compressor conversion involved installing a hydraulically actuated head end variable volume clearance pocket on the 1st stage cylinder along with a pneumatically actuated fixed volume clearance pocket on the head end of the 2nd stage cylinder as shown in Figure 13-33. The control pressure for the hydraulically actuated variable volume unloader was provided by the use of discharge pressure acting on an external accumulator containing hydraulic fluid separated by a bladder. This approach eliminated the need for separate pumps and power requirements. The control pressure for the fixed volume pocket was provide by plant air pressure.

Control algorithms, that were derived from a robust reciprocating compressor performance modeling program developed by the author's company, were programmed into a PLC, which then automatically controlled the unit with the two unloaders. The conversion required about

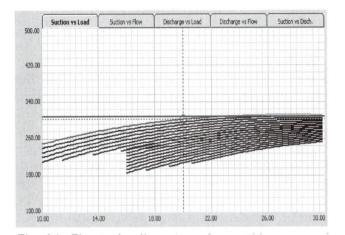

Figure 13-32. Fine unloading steps for a wide range of operating conditions are provided by a combination of fixed and variable unloaders.

Figure 13-33. New automatic clearance pockets added to propane refrigeration compressor.

two days of installation time, and it has been operating continually for more than 3 years.

Case History 16: Pipeline Compressor Automation

A pipeline operator required the reapplication of five large Clark TCV-12 and TLA-6 19.0 in. (483 mm) stroke integral engine compressors. An increased range of pressure ratios required a reduction of existing fixed clearance. The budget was tight and required reusing as much of the existing hardware as possible. The author's company modeled the compressor performance and developed an optimum unloading configuration. The front and rear heads and liner ports were reconfigured to reduce the cylinder fixed clearance. The large 14.0 in. (355.6 mm) diameter cylinders were lined down to 12.5 in. (317.5 mm). The compressor valves were redesigned to maximize the lift area and minimize fixed clearance. A total of 116 fixed volume clearance pocket unloaders having volumes of 150 in3, 300 in3, 600 in3, or1500 in^3 (2.5, 4.9,9.8 or 24.6 L) provided the flexibility necessary to meet the wide range of compression ratios associated with the new application. Novel piston designs permitted the use of existing piston rods while pro-viding piston to rod attachment features that resulted in easier and safer assembly techniques. The revamped units, shown in Figure 13-34, provided the required performance, which was more flexible and reliable than the OEM's offering.

Figure 13-33. One of five large 19.0 in. (483 mm) stroke integral engine compressors that were completely configured for automatic operation with a broad range of pressure ratios.

Case History 17: Replacement of
Variable Speed Engine with Synchronous Electric Drive

A natural gas producer had a 2-throw, 2-stage Superior W-72 reciprocating compressor driven by a 1600 hp (1193 kW), 900 rpm gas engine that had become obsolete, was inefficient and was expensive to maintain. In addition, due to local air quality restrictions, and electric motor drive was considered as a preferred replacement for the engine drive. The primary mode of flow control was via speed control on the engine, which provided a high level of flexibility. However the operator did not want the complexity or cost of a variable frequency electric motor drive. Use of a constant speed, synchronous motor drive, then, appeared to be impractical for the application that had to have a wide range of flow control capability. The author's company was requested to propose unloading equipment that would provide optimal flexibility at reasonable cost. After weighing various options, the preferred solution involved simply replacing the outer heads of each compressor cylinder and adding plug-type suction valve deactivators to the head ends of both cylinders. As shown in Figure 13-35, the new 1st stage head was configured with a custom engineered design having 3 different clearance pockets of 73, 146 and 292 in^3 (1.2, 2.4 and 4.8 L) respectively.

Figure 13-35. Two-throw compressor speed with multiple clearance pockets on each compressor cylinder head end.

The 2nd stage head was configured with a custom engineered design having 2 identical clearance pockets of 170 in^3 (2.8 L). Each pocket was controlled with its own integral pneumatic actuator that was controlled by a PLC control panel mounted on the compressor package. Together with the two sets of head end suction valve deactivators, this arrangement resulted in 31 discrete load steps that provided the very fine control shown in Figure 13-36. As shown, the arrangement actually provided about 36% more turndown capability than the original variable speed drive.

Case History 18: Automation of Field Gas Gathering Compressors

A U.S.A. Appalachian natural gas producer was operating three 2-stage Ajax DPC-600 compressors. Suction pressure swings lead to the units shutting down to prevent overloading, thus reducing each unit's run time and requiring personnel to travel out to the remote units to restart them. The operator therefore had been running the units conser-

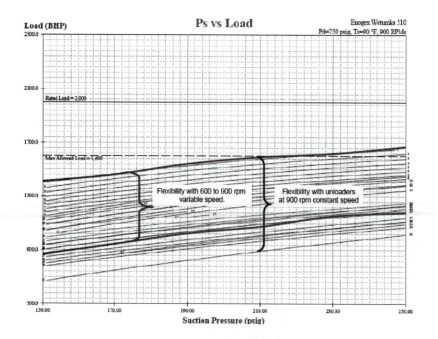

Figure 13-36. The multiple pocket unloaders, together with valve deactivators on the two head ends, provided more flexibility with a synchronous motor drive than the same compressor had with a 33% speed turn down.

vatively in order to accommodate the pressure swings without shutting down, however that resulted in less overall production.

An engineering study was commissioned to evaluate the compressor configuration to identify the best means for operating each unit via automatic control and unloading to cover the entire operating range. Goals were to keep the engines running safely as field condition changed, and also to increase capacity during normal operation to minimize the number of units that had to be operated at any time. Various alternatives were evaluated including a single 685 in³ (11.2 L) fixed volume pocket; parallel 192 and 383 in³ (3.1 and 6.3 L) fixed volume pockets; parallel 114, 228, and 342 in³ (1.9, 3.7 and 5.6 L) fixed volume pockets, and an automatic variable volume pocket. In order to evaluate the alternative solutions, performance was compared at more than 1000 operating points within the operating range noted in Figure 13-38. The highest flow possible at each point was plotted for each case at each condition and the curves were overlayed for comparison. Over

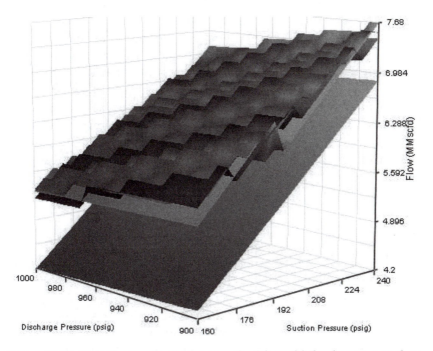

Figure 13-38. Two-throw compressor speed with multiple clearance pockets on each compressor cylinder head end.

Figure 13-39. 3-volume clearance pocket unloader.

the range of 160 to 240 psig (11.0 to 15.5 bar) suction and 900 to 1000 psig (62.1 to 69.0 bar) discharge, the highest average flow was possible with the automatic variable volume pocket, however, the 3-volume pocket (the 2nd from the top surface in Figure 13-38) provided average flow that was within 1% of the variable pocket (top surface). The 3-pocket heads, Figure 13-39, were also slightly lower in cost and significantly easier to control. This solution was also a more practical alternative for a remote field application with limited technical support and maintenance available.

SUMMARY

The list of examples can go on and on, but the foregoing case histories provide a range of good examples. Optimization and revitalization has been successful on compressors used in air drilling, nitrogen, air procession, air separation, chemical processing, field gas gathering, gas boosting, gas lift (EOR), gas transmission, gas injection and withdrawal, refinery, refrigeration and specialty applications. Virtually any size and model can be considered for improvements, whether or not the compressors are currently supported by OEMs or are obsolete. Experience includes Ajax, Alley, Ariel, Chicago Pneumatic, Clark, Cooper-Bessemer, Dresser-Rand, Delaval, Enterprise, Gardner Denver, GE Gemini, Hurricane, Joy, Knight, Neuman Esser, Norwalk, Penn Process, Superior, Thomassen, and Worthington.

Acknowledgement

The author acknowledges the legacy provided by many decades of technical leadership, innovation and dedication of respected industry experts William Hartwick, Richard Deminski, Edward Miller, and Richard Doup, as well as the ongoing contributions of Chad Brahler, Dwayne Hickman, Charles Wiseman, Larry Burnett and their capable teams.

References

1. Brahler, C.D.; Hickman, D.A. & Shade, W.N., Performance Control of Reciprocating Compressors - Devices for Managing Load and Flow, Gas Machinery Conference, Albuquerque, NM, Oct. 7, 2008.
2. Brahler, C.D. & Schadler, H.C., One Compressor OEM May Not Have the Optimum Equipment Selection, Gas Machinery Conference, Dallas, TX, Oct. 1, 2007.
3. Harbert, P.; Hickman, D.; Mathai, G., & Miller, R., An Automation Method for Optimizing and Controlling A Reciprocating Compressor Using Load Step, Speed and Suction Pressure Control, Gas Machinery Conference, Dallas, TX, Oct. 2, 2007.
4. Shade, W.N., Custom Engineered High-Performance Cylinders Revitalize Existing Compressor Assets, *GMC Today*, Oct. 4, 2004.
5. Shade, W.N., DCP Midstream Upgrades LaGloria Gas Plant, COMPRESSORTech-Two, Oct. 2007.
6. Shade, W.N., Modern Replacements for Old Valve-In-Head Compressor Cylinders, *GMC Today*, Oct. 2, 2007.
7. Shade, W.N., Re-Engineering Extends the Utilization of Pipeline Compressor Assets, *GMC Today*, Oct. 3, 2006.

Chapter 14

Piston Rod Run-out is a Key Criterion for Recip Compressors

—Gordon Ruoff

INTRODUCTION

Proper assembly, installation and maintenance are the keys to compressor efficiency and life. And the major key is piston rod run-out. Even though the compressor may have been test run at the factory, piston rod run-out must be rechecked on site as part of the compressor installation. Thereafter, any time the piston or rod is serviced, rod run-out must be checked. This paper established the criterion for rod run-out and was included in API "Standard 618 Reciprocating Compressors for Petroleum, Chemical, and Gas Industry Services", Appendix C.

This article, published on October 12, 1981 was provided courtesy of *Oil & Gas Journal*. The article was written by Gordon Ruoff.

Piston rod run-out is a criterion used by engineers and operating personnel to determine piston-rod alignment in relation to cylinder and crosshead alignment on horizontal reciprocating compressors. Stringent rod run-out requirements are frequently part of the purchase specifications for reciprocating compressors.

Factory witness of a rod run-out check is often a requirement of the inspection of a new compressor. Rod run-out is usually always checked as a part of normal recip compressor maintenance, especially after overhaul and reassembly of the gas ends.

HORIZONAL RUN-OUT

Horizontal rod run-out readings can be used as a direct indication of horizontal alignment from the crosshead through the distance pieces

Figure 14-1.

to the cylinder. Horizontal rod run-out is measured by placing a dial indicator pickup on the side of the rod.

For perfect alignment, the indicator should read zero as the rod is moved through the length of the stroke. Variations of ± 0.00015 in./in. of stroke are usually considered acceptable.

Horizontal rod run-out should be the same whether the unit is cold or hot. Excessive horizontal rod run-out is corrected by realignment of components involved. This may include cylinders, heads, distance pieces, crossheads, crosshead guides, rod, and piston. Squareness of rod threads relative to piston-rod nut threads and face and crosshead threads and face is a critical factor.

VERTICAL RUN-OUT

Normal expected cold vertical rod run-out is not an indication of misalignment. For perfect alignment, it is the result of the difference between the normal cold running clearance of the piston to bore and the crosshead-to-crosshead guide, which in large cylinders causes the piston

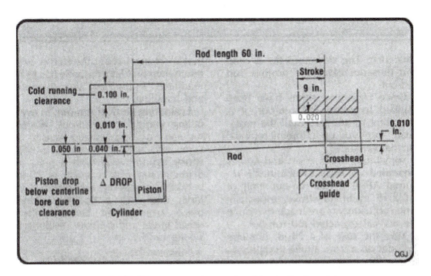

Figure 14-2. Basic geometry of cold vertical rod run-out
(As measured by a dial indicator during a manual bar-over)

centerline to lay below the crosshead centerline. This basic geometry is illustrated in Figure 14-2.

Note that the piston and crosshead centerlines lay below the perfect alignment centerline by one half of the running clearances. In cylinders where the running clearance is greater (or less) than the crosshead running clearance, the piston will lay below (or above) the crosshead centerline by one half the difference in the cold running clearances.

The result is normal critical rod run-out. For purpose of calculating normal vertical rod run-out, this one-half clearance difference is designated as the differential drop (Δ drop).

From Figure 14-3, it can be seen that the vertical rod run-out had a direct relationship between these normal running clearances, Δ drop, rod length, and stroke. This creates a nearly proportional right-triangle situation such that the normal expected cold vertical run-out can be calculated with sufficient accuracy when these values are known.

Vertical rod run-out is measured by placing a dial indicator on the top or bottom of the rod. As the rod is moved through the entire stroke length, it will "run-out" by an amount equal to the ratio of the stroke-to-rod length times the Δ drop, i.e., half the difference between the running clearances (Figure 14-3).

It is generally desirable that the measured vertical cold rod run-out

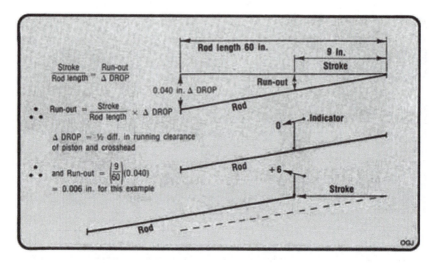

Figure 14-3. Vertical rod run-out relationships.

be within the calculated value (for the involved cylinder and crosshead clearances, rod length, and stroke) by ±0.00015 in./in. of stroke. Any amount of rod run-out outside of these limits may indicate some misalignment.

While it may be thought by some that "zero" vertical cold rod run-out is highly desirable and is a sign of perfection, it may actually be an indication of some misalignment, particularly at the time of new unit assembly where large nonlube cylinders with aluminum pistons and Teflon rider rings are involved. It is necessary, in fact, to consider the run-out at both ambient and normal operating temperatures.

NORMAL RUN-OUT

For large cylinders with aluminum pistons and Teflon rider rings, there can be a significant difference between the normal cold rod run-out and the run-out at normal operating temperatures. This is due to the high expansion rate of the aluminum piston and Teflon rider rings which can make a significant difference in the differential clearance between the piston and the crosshead.

On the other hand, there may be operating conditions with low suction temperatures such that normal operating temperatures may be no greater than the ambient temperature upon which the normal expected cold vertical rod run-out is based. These situations must be given special consideration with the cold rod run-out adjusted accordingly.

ACCEPTABLE VERTICAL RUN-OUT

What is an acceptable vertical rod run-out figure? API-618 2nd Edition requires 0.00015 in./in. of stroke at normal operating temperatures. Although this figure is deemed by most manufacturers as being too stringent, especially where large cylinders are involved, it can usually be accomplished by shim adjustment of the crosshead shoes. Shim adjustment must be made on the determination of both the normal expected cold rod run-out and the expected run-out at the normal operating temperatures.

A thorough study of the figures and readings is required to determine what adjustments, if any, are needed to obtain the desired run-out at operating temperatures. It is felt by some that 0.001 to 0.002 in./in. of stoke may be perfectly acceptable for normal operating vertical run-out, especially where large cylinders are involved. This amount of run-out creates just enough movement in the packing rings to keep them from sticking in cups, especially in dirty gas applications.

Where there is much concern about rod run-out, each application needs to be studied carefully to make the right decision and proper adjustment. There are many compressors operating successfully with the higher rod run-out figures.

MEASURING ROD RUN-OUT

With regard to vertical rod run-out, some clarification of positive and negative magnitude is needed. Rod run-out should always be measured starting with the rod at the extreme end of the stroke.

If the rod is all the way "back," i.e., with the piston at the crank end of the cylinder, and the dial indicator placed on top of the rod at the 12 o'clock position and set at zero as in Figure 14-3, when the rod is stroked "forward," i.e., out toward the head end, the indicator will read positive, indicating that the rod is sloping downward from the crosshead to the piston as shown in the illustration. (If the indicator is placed on the bottom of the rod at the 6 o'clock position, it would read negative.) This is the normal situation encountered where cylinder clearance exceeds crosshead clearance.

If the rod is all the way out (piston to head end) and the indicator is set at zero, then it will read negative as the rod is stroked "back." This negative reading in this direction still indicates the rod is sloping "downward," as does the positive reading obtained when the rod is stroked outward during the reading.

To avoid confusion, it is suggested that all rod run-out readings be taken on a forward stroke with the dial indicator set at zero when the rod is all the way back. On multithrow units, this is important so that relative readings among the throws can be properly compared and evaluated.

CORRECTING VERTICAL RUN-OUT

Vertical rod run-out should never be adjusted or corrected by forcing the cylinder and/or distance piece up or down by use of support adjusting screws. This can put excessive forces and stresses on the components involved. Cylinder alignment and cylinder level at both cold and normal operating temperature conditions should be determined and adjusted so that components will be free of harmful stresses at normal operating temperatures. Then rod run-out should be checked and corrected if necessary.

If alignment is determined to be basically correct, and it is decided that some adjustment to the normal vertical rod run-out is absolutely necessary, this can be done by means of shims under the crosshead shoes, e.g., taking shims from the bottom shoe and placing them under the top shoe drops the centerline further below the perfect alignment centerline while maintaining the same running clearance. This decreases the Δ drop and thus decreases the positive rod run-out.

Once crosshead shims have been shuffled to adjust rod run-out, it is important to always install the crosshead "top" side up following removal for maintenance.

Sometimes much effort and cost is expended uselessly to obtain the required API-618 rod run-out limit of 0.00015 in./in. of stroke, especially when adjustments are made to reduce the normal expected rod run-out.

Take the case of a 24-in. nonlube cylinder on a 9-in. stroke compressor where the normal vertical cold rod run-out would be about 0.006 in. with a 60-in. long rod and about 0.0045 in. with an 80-in. long rod (cylinder running clearance is 0.090 in. and crosshead running clearance is 0.012 in., making Δ drop equal to (0.090-0.012)/2 or 0.039 in.

If rod run-out is adjusted to zero when cold, it will be about a negative 0.003 in. when hot, i.e., it will be laying upward, instead of downward as shown in Figure 14-3.

However, a 24-in. nonlube piston has a rider ring design with about 0.060 in. of allowable wear built into it. Over time, as the rider ring wears, the piston will drop below the centerline of the crosshead by the amount of the allowable wear. Thus, when it is time to replace the rider rings, rod run-out will be about 0.009 in. with the 60-in.-long rod.

If periodic crosshead shim adjustments have been made in an attempt to keep rod run-out at or near zero, then all the shims that were changed must be put back in their original positions. To do this is unnecessary and costly in relation to any benefits gained. Further, vertical rod run-out that is set precisely at the factory with much time and effort usually has to be readjusted in the field after shipment and handling.

EVALUATING VERTICAL RUN-OUT

In evaluating vertical rod run-out, normal rod sag may occasionally have to be taken into account. Consider a 9-in. stroke cylinder end with a 1-3/4 in. diameter rod with a length of 70 in. between crosshead and piston (cylinder end uses a double distance piece arrangement between crosshead guide and cylinder, resulting in a long rod).

If piston clearance is 0.040 in. and crosshead clearance is 0.020 in., the normal expected vertical rod run-out (from the formula of Figure 14-3) is (9/70)(1/2) (0.040-0.020) or about 0.001 in. ±0.0014 in.

Considering normal sag, the maximum deflection is calculated from the formula $(5/384)(WI^3EI)$ for a beam supported at both ends. Assuming the rod to be steel with a weight of 8.17 lb/ft, E (modulus of elasticity) or 30×10^6, and 1 (moment of inertia) of 0.40604; then:

normal sag = $(5/384) \times [(47.7 \times 70^3)/(30 \times 106 \times 0.4606)] = 0.0115$ in.

However, the projected straightline deflection is about 1.5 times the calculated sag, or about 0.023 in. This is the figure that must be used to calculate the expected rod run-out relative to sag conditions (Figure 14-4). The effective rod length is now only 1/2 the full length or 35 in. Since short rods result in greater run-out, the effective run-out due to sag is "amplified."

With this sag condition, and with the piston and crosshead in perfect alignment, i.e., with no Δ drop, when an indicator is placed next to the crosshead, or next to the packing area (closest to the piston as possible) expected rod run-out due to sag will be 9/35 x 0.023, or about 0.006 in. This, of course, is a condition that cannot be readily corrected.

To determine if rod sag is contributing to what appears to be excessive rod run-out, an indicator placed on the rod next to the crosshead will result in a positive reading as the rod is stroked outward toward the headend, while another indicator next to the pressure packing will result in a negative reading during the same stroking.

An indicator placed on the rod at about its midpoint will result in little or no run-out reading.

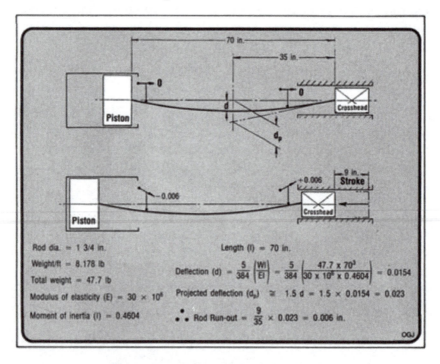

Figure 14-4. Rod run-out relative to sag.

In reality, when the crosshead and piston are subject to rod sag they tend to lay in the crosshead guide and cylinder as shown exaggerated in Figure 5. During the outward stroke when the rod is in compression, the column effect creates even further "sag." Then when the stroke reverses and the rod is in tension it tends to straighten out. This sometimes causes the rod to appear to jump up and down when viewed in operation with the naked eye.

Instruments used to detect rod dynamics during actual operation sometimes record the reversal at the end of each stroke as a significant "jump." If a dial indicator is placed at the lowest point of rod sag during a bar-over test, the difference in deflection from rod compression to rod tension can often be seen, resulting from the frictional drag of the piston in the bore, especially if it is a large nonlube cylinder.

In summary, horizontal rod run-out should ideally be zero, with a variation of no more than ±0.00015 in./in. of stroke. Excessive horizontal rod run-out can be an indication of misalignment.

Normal vertical rod run-out is not an indication of misalignment. Rather it is the result of the difference between the piston and crosshead running clearances, with its magnitude further affected by the length of the piston rod, and by the stroke. It needs to be calculated for each cylinder end. Normal variation from the calculated value should be within ±0.00015 in./in. of stroke.

Some axioms relative to vertical rod run-out are as follows:

- The larger the cylinder and the more the running clearance, the more the run-out.

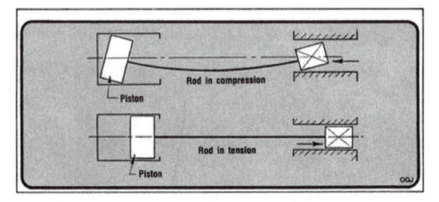

Figure 14-5. Crosshead and piston subjected to rod sag.

- The longer the stroke, the more the run-out.

- The longer the rod, the less the run-out.

- Short rod, long stroke = greater run-out.

- Long rod, short stroke = lesser run-out.

- The larger the cylinder, the longer the stroke, the shorter the rod = greater run-out.

- The smaller the cylinder, the shorter the stroke, the longer the rod = lesser run out.

There are many compressors successfully running with rod run-outs that exceed any of the limits and parameters established above. In some rare cases rod sag may contribute to a misleading excessive run-out reading which cannot be improved by further alignment or crosshead adjustment

Rod run-out that is set precisely to stringent requirements at the factory with the expenditure of much time and effort usually has to be readjusted in the field after shipment, handling, and installation if the same precise run-out is required.

Chapter 15

Effect of Pulsation Bottle Design On the Performance of a Modern Low-speed Gas Transmission Compressor Piston

—Christine M. Gehri
Ralph E. Harris, Ph.D.
Southwest Research Institute

INTRODUCTION

Pulsation bottle design is typically thought of as a means to minimize harmonic responses in the suction and discharge piping. As this chapter shows, there is a lot more to pulsation bottle design. With proper design not only will harmonics be controlled but compressor efficiency will be improved.

This study was presented at the 2003 Gas Machinery Conference by Christine M. Gehri and Ralph E. Harris, Ph.D.

Dynamic pressure drop refers to the calculation of instantaneous pressure drop at each point through the crank rotation. The horsepower cost is computed by integrating the instantaneous pressure drop over a cylinder cycle. If the instantaneous velocity across a square law restriction is much higher than the mean velocity, then the cycle integrated horsepower cost will be much higher than the horsepower cost estimated by mean flow estimates. Field analyses have documented instances where dynamic pressure drop effects have incurred hundreds of excess horsepower and, in some cases, were so high as to inhibit full load operations. In one such case, strong cylinder interactions combined with a significant pressure drop restriction downstream of the compressor cylinder valves resulted in excessive horsepower losses.

This paper presents a case study of a low-speed reciprocating compressor unit installed in a natural gas transmission service. On-site testing following start-up indicated total losses of approximately 55 percent of the indicated horsepower while operating on the bottom load step at rated speed. Interim modifications to the existing primary suction and discharge pulsation bottles were made, which allowed the units to operate much more efficiently through the winter. Redesign of the primary pulsation bottles was necessary to further minimize cylinder interaction and the associated horsepower cost. Full documentation of the design process is presented in this paper, along with supporting field data.

INTRODUCTION

Bottle design is a key component in a successful compressor installation. Traditionally, bottle design focused on controlling pulsation levels, and ultimately vibration, through resonance avoidance. The majority of the low speed integral units installed in the US gas transmission industry were designed using analog technology. The analog approach focused on locating acoustic resonances outside of the normal operating speed range and limiting bottle unbalanced shaking forces to a target value. Empirical loading values, and not stress calculations, were key design parameters. The success of this approach is evident in the continued operation and long life of this compressor fleet. With traditional design tools, the effect of the bottle design on compressor performance is difficult to establish. Because of the linear nature of the solution, the role of valve performance and pulsation levels on the predicted pressure-volume cards was not accurately represented. Moreover, the effects of pressure drop through the bottle system were not reflected in the predicted horsepower, or any equivalent accounting in the design. Following installation, a bottle design was typically viewed as successful if it achieved acceptable vibration and compressor performance, as measured by cylinder performance testing.

It is very difficult to measure the impact of the pulsation bottle design on cylinder performance in the field. Although pulsation levels are easily measured, differential pressure measurements between various locations in the bottle design or between the cylinder nozzle and header are not only unreliable, the values are difficult to interpret with regard to horsepower costs. Pressure drop through the cylinder valves and orifice pressure loss may show up as differential indicated pressure on the pressure-volume card, but through the bottle system, these losses are typically seen as a spread in the cylinder toe pressures. A survey of compressor efficiencies[1] in the installed base of integral units in the US gas transmission industry revealed a significant spread, as shown in Figure 1. This spread in efficiency represents a loss of pipeline capacity as well as excess fuel consumption. Given our improving ability to predict and understand the role of bottle design on compressor performance, a significant component to the observed spread must be attributed to pulsation bottle designs.

The increasing use of high-speed compression in various applications has prompted the development of design tools that better meet the demanding requirements of today's high-speed compressor packages. Pulsation bottle design, in particular, has taken on new challenges. Resonance avoidance on variable high-speed units is much more difficult to achieve, and resonance management is required. High-speed units have a more difficult time achieving traditional compressor performance efficiencies, and pulsation control through pressure drop can drive unit efficiencies down further.

Dynamic pressure drop refers to an increase in static pressure drop through a component or an entire system of pulsation control elements, resulting from pulsating flow. Because pressure drop is proportional to the square of the velocity through an element, dynamic flow results in more pressure drop on the swings above the mean flow than is saved on the portion of flow below the mean flow value. When acoustic resonances are involved, the dynamic flow can be significant and produce surprisingly large horsepower losses. Since the compressor is the prime mover of the gas, these dynamic pressure losses are reflected in the horsepower consumed by the compressor cylinder and seen in the pressure-volume cards. With the ongoing transition to digital acoustic analysis, calculation of dynamic pressure drop can be automated into the design process of future systems. However, the large body of existing compressors may benefit from reevaluation of the pulsation control system, particularly when electric drives are involved or capacity limitations have been met.

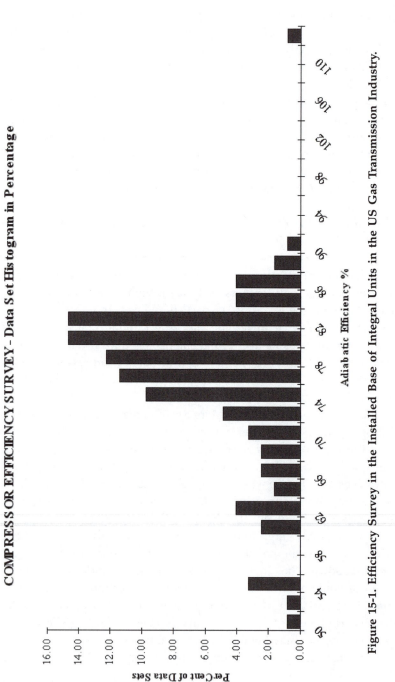

Figure 15-1. Efficiency Survey in the Installed Base of Integral Units in the US Gas Transmission Industry.

CASE STUDY 1

Figure 2 is a photograph of a Cooper GMW330 installed in 1996 at the Enterprise Compressor Station operated by Southern Natural Gas Company. This is one of the last low-speed integrals installed in the US gas transmission industry. The unit is rated at 3,920 HP and has two 21-inch bore cylinders. The cylinders are down connected with the primary suction and discharge bottles located below the compressor deck. The unit operates over a speed range of 255 to 330 rpm, with fixed clearance pockets and a total of 19 load steps. The original primary bottle design, shown in Figure 3, used a baffled, center fed bottle with a perforated choke to reduced second order pulsations. This is one of three identical units installed as part of an early horsepower replacement project. Following installation, the unit could not run off of the bottom load step without exceeding the rated engine torque. Figure 4 presents the cylinder pressure data from Cylinder 2 at 330 rpm. As shown, compressor efficiency (based on ideal compressor horsepower between toe pressures)

Figure 15-2. Cooper GMW330 with 21-inch Bore Cylinders.

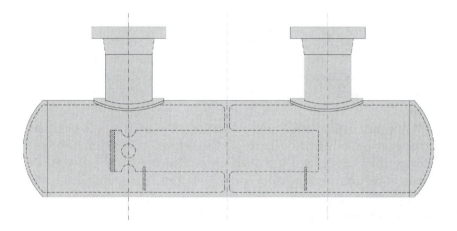

Figure 15-3. Cooper GMW330 Original Bottle Design – 1996.

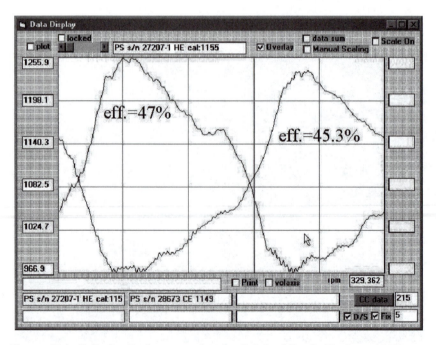

Figure 15-4. Cylinder Performance for the Original Bottle Design at 330 rpm.

was on the order of 46 percent. Note that the differential indicated power (DIP) above and below toe pressures is consistent with very large valve losses, or perhaps nozzle orifice losses. Simultaneous measurements of pressure behind the suction and discharge valves, overlaid with cylinder pressure data clearly indicated that high valve loss was not the root cause of the excessive losses.

Using the measured dynamic pressures, combined with the acoustic model developed for the system, an estimate of the static and dynamic losses through the choke tube was made. Figure 5 illustrates the nature of these calculations. The flow through the perforated choke tube is first established. From this flow estimate, the pressure drop is estimated using the static loss coefficient. Using the mean flow and static loss coefficient, the estimated pressure drop is 2.8 psi. Using the same loss coefficient and the dynamic flow, the average estimated pressure drop is 17 psi. Using mean flow and mean pressure drop estimates, the cost of the perforated choke is estimated to be 30 HP. Using the dynamic flow and dynamic pressure drop estimates, the cost of the perforated choke is estimated at 600 HP. Field modifications of both the suction and discharge bottles consisted of shortening the choke tubes and removing the perforated section, as shown in Figure 6. Shortening the choke resulted in significantly raising the choke's resonant frequency, an acoustic mode involving strong cylinder-to-cylinder interactions.

Figure 7 presents the cylinder pressure data following modifications to the suction and discharge primary bottles. Based on the improved cylinder efficiency, 440 of the estimated 600 HP were recovered. This particular station typically operates at a low ratio, which makes it difficult to achieve low pulsation levels and consistently high cylinder efficiencies. Based on data acquired at various speeds and operating conditions, a simple relationship between pulsation driven cylinder losses and combined suction and discharge nozzle pulsations was developed. As shown in Figure 8, combined suction and discharge nozzle pulsations of less than 40 psi peak-to-peak would be required to drive losses to less than 50 HP per cylinder end.

The units operated with this bottle configuration until 2002, when interest was expressed to recover more of the lost horsepower through the pulsation control system. A bottle system was designed that would completely isolate cylinder-to-cylinder interactions. A goal of the design was to use the existing secondary bottle located outside of the compressor building. Figure 9 presents a schematic of the new design approach.

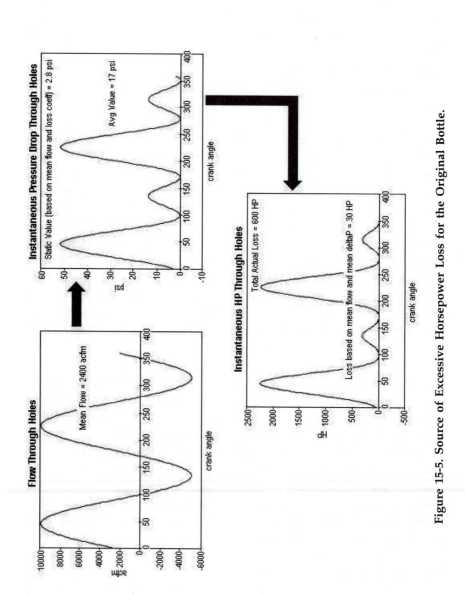

Figure 15-5. Source of Excessive Horsepower Loss for the Original Bottle.

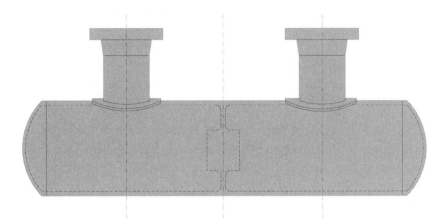

Figure 15-6. Cooper GMW330 Modified Bottle Design – 1997.

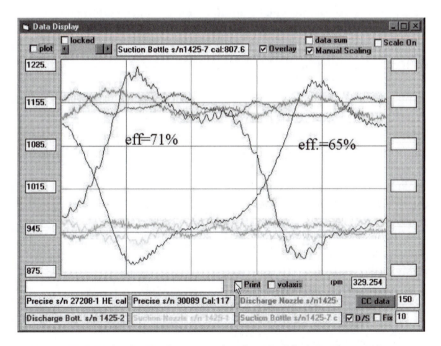

Figure 15-7. Cylinder Performance for the Modified Bottle at 330 rpm.

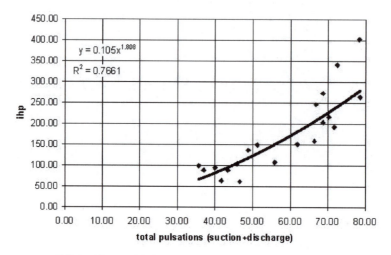

Pulsation Dip HP versus total pulsations

Figure 15-8. Effect of Residual Pulsations on Cylinder Performance for the Cooper GMW330 Unit.

Both cylinders are acoustically isolated from each other by using a long choke to connect the primary bottle chamber for each cylinder to the external bottle. Figures 10 and 11 present pictures of the dual choke configuration leaving the primary bottle and the small bottle used to recombine the flow prior to entering the exterior secondary bottle. Figure 12 presents cylinder pressure traces following modifications to the bottles. Table 1 summarizes the current performance of the unit. Compressor performance is now up to 92 percent at comparable operating conditions.

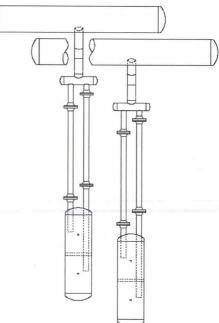

Figure 15-9. Current Bottle Layout for the Cooper GMW330 Unit.

Figure 15-10. Dual Choke Configuration.

Figure 15-11. Special Fitting for Dual Chokes.

A final concern expressed by the unit operators was an apparent variation in discharge pressure between the three units. The center unit (Unit 12) displayed higher discharge pressure than Units 11 and 13. The pressure sensors for all three units were located in the center of the choke tubes between the primary and secondary bottles. Again, dynamic flow losses were thought to be driving the variation in discharge pressures. Dynamic pressure data were acquired in the center of the suction and discharge choke tube as a function of unit speed. Figure 13 presents the results and clearly shows the choke tube resonance excited at 300 rpm. Figure 14 presents the static discharge pressure recorded at the same location. Clearly, the dynamic response is driving the increase in apparent static pressure.

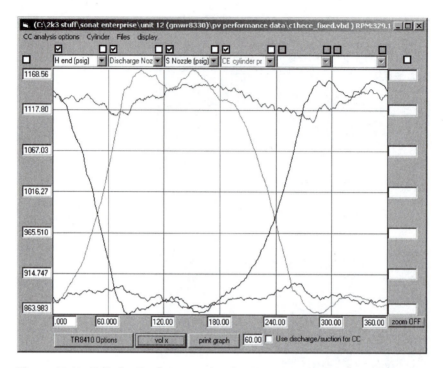

Figure 15-12. Cylinder Performance for the Current Bottle Layout at 330 rpm.

Table 15-1. Performance Summary for Current GMW330 Bottle Layout.

Cylinder End	HE No. 1	CE No. 1	HE No. 2	CE No. 2
Total Cylinder HP	1852.2		2008.5	
Discharge DIP	2	8.4	2.3	7.1
Suction DIP	3.8	1.5	4.3	1.1
Total Discharge DIP	5.2		4.8	
Total Suction DIP	2.6		2.6	
Cylinder Efficiency	92.2		92.6	
Cylinder MMSCFD	71.1	62.7	68.5	67.9
Cylinder Ratio	1.28		1.28	
Total Unit IHP	3860.7			
Total Unit Flow (MMSCFD)	270.2			
Compressor IHP/MMSCFD	14.3			

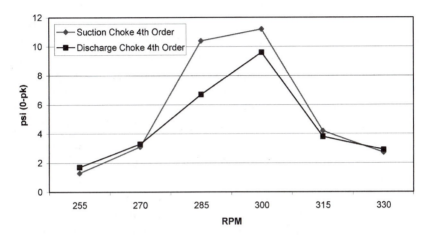

Figure 15-13. Unit 11 Choke 4th Order Dynamic Pressure Versus RPM.

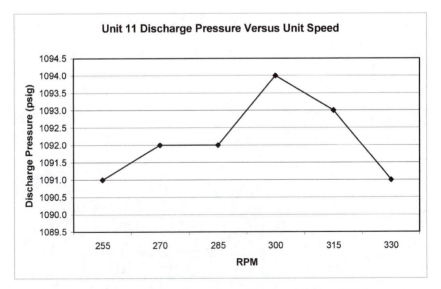

Figure 15-14. Unit 11 Discharge Pressure Versus RPM.

CASE STUDY 2

Figure 15 is a photograph of a Dresser-Rand HHE reciprocating compressor unit operating in a storage application. There are three identical units at this station, installed in the late 1970's. The units are separable electric driven compressors operating at 327 rpm and rated at 6,000 HP. The units can be operated in either a single or two-stage injection/withdrawal service with discharge pressures up to about 3,000 psig or a single stage withdrawal service with discharge pressures of about 900 psig. The operating company plans to install a fourth identical unit and requested a reevaluation of the existing bottle designs. SwRI's digital Interactive Pulsation-Performance Simulation[2,3] (IPPS) code was used to evaluate the current bottle design. The IPPS code automates the calculation of dynamic flow losses at pressure drop elements and the analysis indicated excessive losses on both the first and second stages (Figure 16). The primary discharge bottles are three-chamber with chokes in series, as shown in Figure 17. Based on the model predictions, the excess losses appear to be associated with dynamic flow through the exit choke between Cylinders 2 and 4. This choke tube carries the flow of two cylinders, and is sized such that the velocity is greater than the

Figure 15-15. Dresser-Rand HHE Low-Speed Separable Unit.

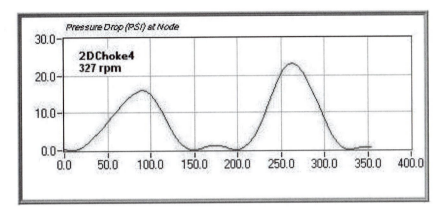

Figure 15-16. IPPS Pressure Drop Predictions for Second-stage Discharge Choke.

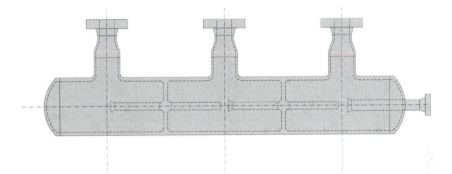

Figure 15-17. Existing Bottle Configurations for HHE Units.

flow through the choke connecting the Cylinder 4 and Cylinder 6 bottle chambers.

As part of an on going GMRC research effort to evaluate and enhance the SwRI/GMRC design tools, a field test effort was completed to document system performance. Cylinder performance data were used to quantify unit losses. Figure 18 presents the head end and crank end traces for Cylinder 2. Table 2 summarizes the unit's cylinder efficiency. Note that, as predicted, there is excess loss of horsepower for Cylinders 2 and 4. Predicted losses are approximately 160 HP on the high-pressure side and 60 HP on the low-pressure side. Combined, the total losses are approximately 4 percent of rated horsepower at the conditions analyzed.

We anticipate that the losses will be greater under different operating conditions. Modifications were developed that should minimize the horsepower cost. At this time, we are recommending that the exit choke tubes be shortened to shift the acoustic response frequency. For electric drive units, the potential savings in horsepower is magnified by demand charge costs, which can significantly alter the economics of unit operation.

SUMMARY

There is a wide range of compressor efficiencies in the installed base of horsepower in the US gas transmission industry. Traditional bottle design methods typically did not quantify the impact of pulsation control on compressor performance. This chapter has shown how dynamic flow losses can significantly reduce compressor efficiencies. Based on our current understanding of the relationship between pulsa-

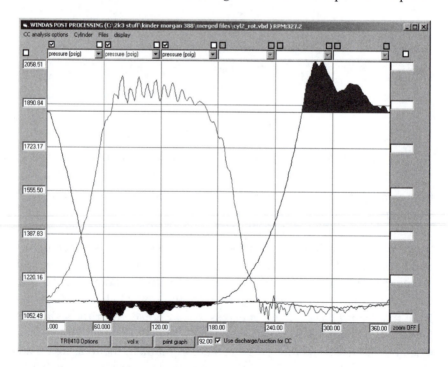

Figure 15-18. Cylinder 2 Performance for Existing Bottle Design.

Table 15-2. Cylinder Efficiency for Existing HHE Bottle Design.

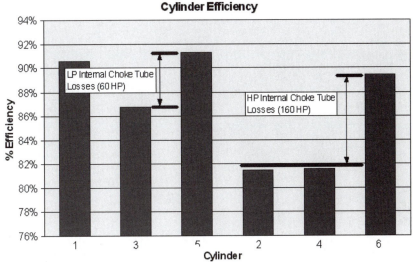

tion control and compressor performance and our continually developing predictive capabilities, we suspect that there are many existing compressor installations that would benefit from a reevaluation of the bottle design. In some cases, even marginal increases in compressor performance can translate directly to significant savings in operating costs due to decreases in fuel consumption and increases in system capacity.

References
1.   Berry, Renee, "Gas Transmission Compressor Efficiency Survey." GMRC Research Project Conducted by Southwest Research Institute, 1995.
2.   Gehri, Christine and Harris, Ralph, Ph.D., "High Speed Reciprocating Compressors – The Importance of Interactive Modeling." Presented at the Gas Machinery Conference, Houston, TX, October 1999.
3,   Gehri, Christine and Harris, Ralph, Ph.D., "Technology for the Design and Evaluation of High-Speed Reciprocating Compressor Installations." Presented at the European Forum for Reciprocating Compressors, Dresden, Germany, November 1999.

Chapter 16

Resolution of a Compressor Valve Failure: A Case Study

—Steve Chaykosky
Dresser-Rand Company
Painted Post, NY

This study addresses the various components of reciprocating compressor valves (i.e., valve plate type, valve plate lift, valve plate impact velocities) the effects of particulates in the gas and how each influences compressor performance and overall operation.

This study has been provided courtesy of Dresser Rand, Inc., the owner of this material. Furthermore, Dresser Rand, Inc., has granted copyright permission for the use of this study in this book.

ABSTRACT

The commissioning date of three critical path ethylene compressors at a Belgian chemical plant was in jeopardy because the original compressor valves failed within hours of start-up on the final stage of compression. The valve supplier attempted to resolve the failures twice, without success. The failed concentric ring valve design was evaluated by the compressor OEM and determined to have marginal reliability characteristics. A different type of compressor valve, the ported plate valve, was proposed as a potential solution. Although a slight decrease in compressor efficiency was expected with the alternative valve technology, the client deemed this an acceptable compromise to meet his production schedule. The new valves were installed in April 1999 and continue to run without problems.

INTRODUCTION

Valve problems are reported as the primary reason for reciprocating compressor shutdowns. The anecdotal evidence is confirmed by a recent survey of hydrogen compressors, which finds that 36% of unscheduled downtime is due to valve failures.

Compressor valves fail for a variety of reasons, which can generally be divided into design factors and operational factors. Design factors include application and selection errors. Operational factors are related to the process gas conditions and to how the compressor is run and maintained.

Discovering root causes of failure depends on a full exchange of information between the valve manufacturer and his client. The client must be willing to provide complete operational data and the valve manufacturer must be willing to review and modify his valve design. Since each reciprocating compressor is unique in the way it is operated and serviced, the circumstances surrounding each case of valve failure and its resolution are also unique.

This case study illustrates how a valve failure can be attributed both to valve design considerations and to operational problems. As the case is discussed, emphasis will be placed on the analysis of valve dynamics, the relationship of valve lift to valve life and compressor efficiency, and elements of valve failure analysis.

The client reported that failures of the 3rd stage discharge valves occurred within hours of achieving full pressure load on 2 of 3 newly commissioned ethylene compressors. The failed valve components were multiple concentric rings made of PEEK, a high-strength thermoplastic.

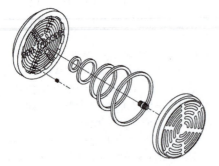

Figure 16-1. Concentric Ring Valve

Figure 16-2. Ported Plate Valve

Figure 16-1 shows an example of a concentric ring valve. The original valve supplier reported that excessive stresses were imparted in the rings due to unexpected radial, fluttery movement. Their response was to design and manufacture a different type of concentric ring element to better resist the lateral motion.

As the compressor OEM and a valve manufacturer, Dresser-Rand was solicited for comments but did not have enough information to perform a formal failure analysis. However, plant startup was imminent, so the client contracted us to provide valve hardware in case the original supplier's modified valves were unsuccessful. We had the freedom to provide whatever valve style we believed would work best. By day's end, a ported plate valve was selected, a dynamic valve analysis (DVA) was performed based on the reported field conditions, new valve drawings and programs for the N/C (numerically controlled) machine tools were created, and the manufacturing floor was alerted to expect a crisis order. Figure 16-2 shows a ported plate valve.

Two days later, the client contacted us in the morning to say that the original 3rd stage suction ring valves had also failed and wanted to know if ported plate suction valves could be delivered with the new discharge valves. Suction and discharge valves for all 3 compressors were manufactured and shipped that afternoon. A set of ported plate valves was installed in one compressor immediately upon receipt.

Ten days after the problem arose, the original valve supplier transmitted simulated P-V (pressure-volume) indicator cards and valve motion diagrams of the failed valves for our evaluation. The data suggested that the original valves were closing on time. However, the original valve supplier's modified valves continued to fail.

By the second week, a second set of ported plate valves were fitted to another compressor, and a site visit by Dresser-Rand personnel yielded the first reports of oil sticktion problems and of particulates in the gas stream.

Since the 3rd stage discharge valves failed first, the client initially believed debris and/or liquids could be eliminated as the problem source because the 3rd stage suction valves, which are upstream of the discharge valves, remained intact. However, the 3rd stage suction valves failed the day after the discharge valves failed, and the client eventually acknowledged that there was debris in the gas stream.

The original valve supplier explained that the radial movement of the concentric rings was caused by valve flutter, which occurred because the compressor did not achieve full pressure fast enough. Flutter is a phenomenon in which the moving elements of a valve open and close several times throughout one compression cycle. To minimize flutter, an accelerated load profile was attempted, which brought the compressor to full pressure faster. However, the 3rd stage discharge valve rings failed again after just a few hours at full pressure.

When asked to comment on the cause of failure, we could only conjecture because failed hardware was not available to evaluate, a dynamics study of the original valve design was not initially provided, and not all of the operational facts were known.

Proprietary DVA software is the tool that was used to generate simulated lift diagrams of the concentric ring and ported plate valves. Since the failed concentric ring valve data was not immediately available, we estimated the configuration of rings and springs in order to perform a DVA study. Figure 16-3 shows the estimated head end discharge valve motion of the concentric rings. Note that all of the rings close before reaching the top dead center (TDC) crank angle of 360°. This is important because it means that the original rings should have closed on time. The original valve supplier's simulated lift diagrams were transmitted to us later, and their valve motion study also indicated on-time closing. Since we developed the ported plate valve data for the DVA, there was a high degree of confidence in the valve motion simulation, and Figure 16-4 shows that the ported plates do indeed close well before TDC.

Lift diagrams illustrate the importance of proper valve closing but do not indicate whether the impact velocities of the moving elements (plates, rings) are acceptable. DVA software calculates impact velocity, but this parameter is useful only if the application limits of each valve type are well known. Different plate and ring geometries have different

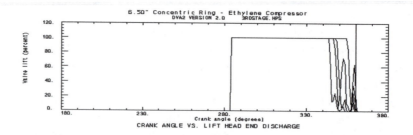

Figure 16-3. Lift Diagram Concentric Ring Valve Head End Discharge Valve

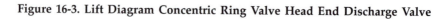

allowable impact velocities, which are experientially derived.

Ported plate valves were selected for this high-pressure ethylene application with the assumption that oil sticktion and excessive plate impacts factored into the failures. Confirmed reports of oil sticktion and debris in the gas stream were received almost 2 weeks after the initial failures occurred.

The emergent nature of this field problem forced a rapid evaluation of potential valve designs. The final selection was certainly based on parts availability and shop capacity, but the foremost customer requirement was also achieved: reliability.

The ported plate valve has less flow area than the original failed ring valve because the lift is lower. Although the client preferred not to have compressor efficiency degraded by a valve change, he deemed it an acceptable compromise. After all, a compressor that shuts down on valve failures has zero efficiency. The effective flow area (EFA) of the ported plate valve is 41% less than the reported EFA of the failed concentric ring valve. A valve's EFA is determined by laboratory flow testing and is typically 40% to 70% of the valve lift area. Valve lift area is a calculated quantity based merely on valve geometry and is open to interpretation by different valve manufacturers. Therefore, EFA is the relevant flow area for the valve designer, not lift area.

The original valve supplier's 2nd modification of the concentric ring valves involved lowering the lifts. The ported plate inlet valves have 17% less flow area than the ring valves with reduced lift; the ported plate discharge valves 29% less. Although the EFAs of the ported plate valves are less than the modified ring valves', an October 1999 monitoring report based on Recip-Trap measurements states that the 3rd stage valve losses were "very low," as were the plate impacts. Simulations

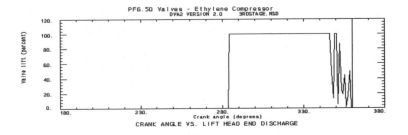

Figure 16-4. Lift Diagram Ported Plate Valve Head End Discharge Valve

predicted a 4.5% increase in 3rd stage power consumption with ported plate valves installed, but the measured losses were actually much less. This points out a common fallacy in calculating a cylinder's power losses. Algorithms that only consider the valve losses and ignore pressure losses over the cylinder passages, cylinder ports, and valve cages are not comprehensive. All these losses are additive. Therefore, a reduction just in valve flow area does not necessarily translate into a proportional power increase and consequent efficiency decrease.

A valve with very low pressure loss appears very efficient on paper. Even when proper springs are chosen to close the valve, there may not be enough pressure drop to hold the moving elements completely open, which will cause flutter. Flutter decreases the effective lift and therefore increases the predicted power losses. Flutter can be eliminated with lighter springs, but then there is a risk of late valve closing, which will reduce cylinder capacity and cause the moving elements to slam shut. If the gas forces slam a valve closed, the moving elements could fracture from the associated high closing impact velocities.

FAILURE ANALYSIS

Design factors working against reliability of the original valve design:

1. *High valve lift.* Lift is simply the distance traveled by the valve's moving elements from the fully closed position to the fully open position. The higher the lift, the higher the flow area of the valve. The higher the valve's flow area, the lower the pressure drop across the valve. The lower the pressure drop across the valve, the lower the horsepower consumed by the compressor's driver (motor, engine, or turbine.) The lower the power consumption, the more efficient the compressor, and the happier the client. By such transitive logic, the client should be happier with higher valve lift, but not in this case. The lift of the original failed concentric ring valve was .079 inches, whereas the lift of the ported plate valve that has been running problem-free for over a year is .040 inches. The lift chosen by the valve designer depends on the valve type and the compressor operating conditions. Different valve styles have different flow characteristics. Ported plate valves have excellent

flow areas at relatively low lifts, but there are quickly diminishing returns as lift is incrementally increased, and there is a certain lift beyond which no more flow area can be obtained. Concentric ring valves are often used at higher lifts than ported plate valves. At the upper end of their useful lift range, concentric ring valves have equivalent or slightly higher flow areas than ported plates, but the impact velocities of the rings at those higher lifts will exceed those of the ported plates. Poppet valves have yet more flow area than concentric ring valves when used at their highest practical lifts, but the poppet's impact velocities may be markedly higher than the rings'. Therefore, choosing the appropriate lift for a given application depends on knowing the impact velocities that different valve types can withstand. In this case, a .040 inch lift ported plate valve works well at the 3rd stage conditions, yet a .079 inch lift concentric ring valve proved unreliable and not measurably more efficient.

2. *Excessive impact velocities.* The laws of physics rarely allow something for nothing, and the price to pay for higher lift is higher impact velocities of the valve's moving elements. Since increased durability of the moving elements depends on relatively low impact velocities and therefore low lifts, the relationship between compressor efficiency and valve reliability is inverse. Reliability and efficiency are competing properties of compressor valves, and an acceptable compromise often means sacrificing efficiency for reliability. In this case, the original valve supplier did not divulge the predicted impact velocities of the concentric rings that failed. However, our estimation of the valve motion indicates that the opening impacts of the original .079 inch lift concentric ring discharge valves may have been 40% higher than the opening impacts of the .040 inch lift ported plate valves, and the closing impacts may have been 38% higher. Lift is not the only factor that determines impact velocity. Compressor speed, cylinder pressure, valve springing, and sticktion are also contributors. Impact velocity increases with higher speeds, higher cylinder pressures, heavier spring rates, and increased sticktion. In this case, the compressor speed of 328 rpm is considered low, but the cylinder pressures of 1000 psi inlet and 3200 psi discharge are relatively high, and the increased 3rd stage oil viscosity and lubrication rate contributed to

a relatively high sticktion level.

3.   *Lateral ring motion at off-design condition* (reduced pressure break-in phase.) Although a compressor valve is often able to work over a range of operating conditions, it is optimized for a single design case, i.e. a single combination of pressures, temperatures, speed, cylinder clearance, gas molecular weight, lubrication rate, and pulsation level. When one of these parameters changes, the valve necessarily runs at an off-design condition, and proper timing of the opening and closing events is disrupted. Conditions of service that are slightly off-design are usually tolerated, but major changes to any of the parameters can pose a severe detriment to the valve motion and sometimes result in failures of springs and moving elements. In this case, the original valve supplier blamed the fluttery, non-synchronous movement of the rings on a slow rate of pressurization of the 3rd stage. The supplier stated that the rings were very sensitive to sticktion effects and special start-up conditions and subsequently redesigned the valve a 3rd time with radiused rings of a different thermoplastic material to fight the sticktion problem and the high impact loading. However, the redesigned concentric ring valves were not installed, so it is not known whether this would have solved the failures.

As the moving elements of a valve open and close, they never move perfectly perpendicular to both stopping surfaces. The plates and rings always wobble before they momentarily rest against the seats and stopplates. Consequently, the initial opening and closing impacts always occur on an outer edge of the plate or ring, which is usually where plate or ring fractures initiate. The ported plate that solved this field problem employs scalloped edges on the outer periphery. Since the plate tips as it moves off the seat and stopplate, the initial impacts will be forced to one of the scalloped edges. Figure 16-2 illustrates the scalloped plate, which is a patented design. Each scallop provides a region of higher cross-sectional area to better absorb the opening and closing impact velocities. Since this increased area can better resist impact stresses, the likelihood of plate failures is reduced. The ported plate is also precision-guided with a center ring and with rollpins to minimize lateral movement. In this case, the plate valve design, which consists of a single ported PEEK plate, resists non-uniform, lateral motion better than the concentric rings.

Client requested the original valve type that failed. Many clients are accustomed to using a particular style of valve in their compressors, and they often develop a comfort level with a given valve type. Preferences are usually experience-based, but price has also been known to enter into the specification process. While customer input is welcomed and encouraged, the valve supplier must know his product's application limits, which are primarily based on pressure, temperature, and compressor speed. Valve designs are customized to satisfy specific operating conditions, and not all valve types will work for all conditions, so it is the valve designer's responsibility to select and advocate the appropriate hardware. While preferences should be honored whenever possible, clients are advised not to dictate final valve design.

Operational factors working against valve reliability:

1. *Particulate ingestion.* The suction strainers on all 3 stages were regularly plugged with small metallic particles. The 2nd & 3rd stage accumulations were associated with cylinder lube oil holding the particles together. The source of the particulates, which were reported to be a reality of the plant's gas process for years, was not divulged. Incompressible substances in the gas stream can become lodged between a moving element (plate or ring) and the stationary parts (seat and stopplate) of a compressor valve. The plate may become overstressed at the point where it hinges on the foreign particles and fail. Plastic plates, such as PEEK or Nylon, will often embed metallic debris and continue to operate properly, but the embedded metal fragments can prematurely wear the sealing surface of the valve seat.

2. *Oil sticktion.* The 3rd stage cylinder lubrication system is independent of the first two stages, which is atypical for reciprocating compressors. The 3rd stage uses a slightly different blend of oil and feeds it at a higher rate than the lower stages. The 3rd stage pressures and temperatures are 1000 to 3200 psi and 95° to 225° F, respectively. Ethylene evidently takes oil into solution at this pressure and temperature combination, so an adhesive agent was added to the 3rd

stage lubricating oil to ensure that it would stick to the cylinder bore, piston rings, and piston rod packing. The additive also caused the oil to stick to the moving parts of the valve. Sticktion is defined as the viscous adhesion of oil between the moving and stationary parts of a valve, and is detrimental to valve reliability in two ways. First, sticktion inhibits plate opening, causing the plate to slam open when the pressure force exceeds the sticktion force, which can cause damage not only to the plate but to the underlying springs. Second, sticktion delays plate closing, which elevates the plate's closing impact velocity. If the plate closes so late that there is backflow through the valve, then there is not only a greater potential for plate damage but there is also an actual loss of cylinder capacity. In this case study, the high oil viscosity and increased lubrication rate contributed to the sticktion effects.

CONCLUSIONS

1. The reasons for valve failures can be divided into design problems and operational problems.
2. Valve lift, plate impact velocity, operating at off-design conditions, valve flutter, oil sticktion, and liquids and debris in the gas stream are factors to consider when evaluating valve failures.
3. There are several different types of compressor valves, each of which has application limits, so a good understanding of how to reliably apply each type is paramount.
4. Successful resolution of valve failures often means having to employ new or redesigned valves that are less efficient. Sacrificing some compressor efficiency for increased valve durability is often an acceptable compromise.
5. Timely resolution of valve failures depends on accurately analyzing complete operational data and any available failed hardware.

Acknowledgements

The author wishes to thank Dr. Derek Woollatt for his thoughtful comments and suggestions on the technical content. Thanks are also due to Anne Lacey for her insightful review. Finally, thanks to Dresser-Rand Company for the opportunity to present this case study.

Appendix A1

Compressor Manufacturers

The following tables list compressor equipment manufacturers and compressors by model. The list is divided into two categories: dynamic compressors and positive displacement compressors. The information provided is instrumental in identifying units with comparable capability.

Note that most positive displacement compressors are packaged by independent companies. These companies package a compressor with a driver per the purchaser specifications (included in the packager's scope of supply are suction & discharge vessels, valves and controls).

This information, provided courtesy of COMPRESSORTech[Two] magazine "Compressor Technology Sourcing Supplement" published by Diesel & Gas Turbine Publications. Current compressor information, as well as Natural Gas Engines, Mechanical Drive Gas Turbines, Electric Motors, Variable Speed Drives, and Mechanical Drive Steam Turbines can be found at www.compressortech2 or www.CTSSNet.net.

Compressor specifications in Appendix XX From the 2009 Compressor Technology Sourcing Supplement, courtesy COMPRESSORTechTwo magazine, published by Diesel & Gas Turbine Publications. Current compressor information can be found at www.compressortech2 or www.CTSSNet.net.

Dynamic Compressor Specifications

Manufacturer	Model Designation	Axial: Multi Stage	Axial: Fixed Stator Vanes	Axial: Variable Stator Vanes	Radial: Single Stage	Radial: Multiple Stage	Radial: Horizontally Split	Radial: Vertically Split	Radial: Integral Gear	Thermal: Integral Electric	Thermal: Single Stage	Thermal: Multiple Stage	OF = Oil Free / OI = Oil Injected	Inlet Flow min acfm (m3/min)	Inlet Flow max acfm (m3/min)	MAWP psig	MAWP bar	Compression Ratio (per stage)	Max Input HP	Max Input kW	Speed_min RPM	Speed_max RPM
Atlas Copco Energas GmbH	GT-Serie				x	x			x				OF	175 (5)	141,000 (4,150)	1,305	90	4	26,820	20,000	1,000	50,000
Atlas Copco Energas GmbH	T-Serie				x	x			x				OF	2,760 (80)	141,000 (4,150)	725	50	4	26,820	20,000	1,000	18,000
Atlas Copco Energas GmbH	H-Serie				x				x				OF	4,950 (140)	24,000 (680)	406	28	2.8	5,360	4,000	16,000	40,000
Atlas Copco Energas GmbH	ZH-Serie				x				x				OF	2,300 (65)	9,200 (260)	145	10	2.8	2,150	1,600	26,400	50,000
Cameron's Compression Systems Group	TA-2000					x			x				OF		1,610 (46)	150	10.3	1.6	350	260		3,600
Cameron's Compression Systems Group	TA-2020					x			x				OF		1,940 (55)	150	10.3	1.6	400	300		3,600
Cameron's Compression Systems Group	TA-3000					x			x				OF		3,360 (112)	150	10.3	1.6	800	597		3,600
Cameron's Compression Systems Group	TA-6000					x			x				OF		8,080 (229)	150	10.3	1.6	1,700	1,270		3,600
Cameron's Compression Systems Group	TA-9000					x			x				OF		11,800 (334)	150	10.3	1.6	2,250	1,680		1,800
Cameron's Compression Systems Group	TA-11000					x			x				OF		12,000 (337)	250	17	1.6	2,500	1,900		3,600
Cameron's Compression Systems Group	TA-20000					x			x				OF		21,000 (595)	510	35	1.6	4,750	3,540		3,600
Cameron's Compression Systems Group	MSG-Alpha					x			x				OF		3,800 (108)	1,200	82.5	1.6	1,250	932		3,600
Cameron's Compression Systems Group	MSG-2					x			x				OF		4,500 (128)	1,125	77.5	1.6	1,850	1,370		3,600
Cameron's Compression Systems Group	MSG-3					x			x				OF		7,000 (198)	1,125	77	1.6	5,100	3,800		3,600
Cameron's Compression Systems Group	MSG-4					x			x				OF		14,000 (397)	725	51.5	1.6	7,000	5,220		3,600
Cameron's Compression Systems Group	MSG-8					x			x				OF		30,000 (849)	725	50	1.6	11,500	8,580		1,800
Cameron's Compression Systems Group	MSG-12					x			x				OF		46,000 (1,303)	687	50	1.6	17,000	12,650		1,800
Dresser-Rand	Pipeline POI					x	x						OF		60,000 (1,700)	1,500	103	12	50,000	37,285		14,500
Dresser-Rand	DATUM					x		x					OF		230,000 (6,513)	10,500	725	1.2	120,000	89,484		27,000
Dresser-Rand	DATUM P					x		x					OF		59,000 (1,671)	3,000	206		120,000	89,484		17,000
Dresser-Rand	DATUM C					x		x					OF		59,000 (1,671)	3,000	206		27,000	20,000		17,000
Dresser-Rand	DATUM ICS					x		x					OF		20,100 (570)	3,000	206		20,100	15,000		17,500
Dresser-Rand	Centrifugal					x	x						OF		230,000 (6,513)	10,500	725		125,000	89,484		38,000
Dresser-Rand	Axial	x	x										OF	40,000 (1,133)	875,000 (8,580)	80	5		125,000	93,000		8,000
Elliott Company	A	x	x										OF	40,000	396,000 (11,200)	80	5.5		120,000	89,500		7,200
Elliott Company	M					x	x						OF		320,000 (9,065)	1,000	69		120,000	89,500	2,680	25,000
Elliott Company	MB					x		x					OF		154,000 (4,360)	10,000	690		120,000	89,500		25,000
Elliott Company	P-LINE					x	x						OF		190,000 (5,380)	90	6.2		10,000	7,460		15,000
Elliott Company	PH-LINE					x		x					OF		45,000 (1,275)	900	62		15,000	11,200		15,000
Elliott Company	TC					x			x				OF		56,500 (1,600)	1,450	100		40,200	30,000		18,500
F S Elliott	PAP Plus				x				x				OF	868	30,600		1.5	12	3,500	2,610		3,600
Garo	VC					x					x		OF	-17	-833		90	1.2		400	3,000	6,000
Garo	VAP					x					x		OF	-17	-833		90	1.2		2,000	3,000	6,000
Garo	CC (Galileo)					x						x	OF	-17	-833		90	3		2,000	4,000	42,000
GE Oil & Gas AC	VH					x		x					OF		60,000 (1,700)	5,000	345		30,000	22,000	3,000	17,000
GE Oil & Gas AC	V					x	x						OF		100,000 (2,830)	700	48		30,000	22,000	3,000	15,000
GE Oil & Gas AC	D					x	x						OF		260,000 (7,370)	150	10					
GE Oil & Gas AC	DH					x		x					OF		50,000 (1,420)	750	52					
GE Oil & Gas Nuovo Pignone	AN (Air Service)					x		x					OF	58,860 (1,667)	353,150 (10,000)	360	25	1.3	20,400	15,000	3,000	17,000
GE Oil & Gas Nuovo Pignone	AN (LNG Service)					x		x					OF	58,860 (1,667)	178,550 (5,000)	360	25	1.3	95,200	70,000	3,000	8,000
GE Oil & Gas Nuovo Pignone	BCL-HP (>350bara)					x		x					OF		6,003 (170)	14,500	1,000.00	1.3	95,200	70,000	7,000	8,000
GE Oil & Gas Nuovo Pignone	BCL-LP/MP (<350 bara)					x		x					OF		29,430 (833)	5,075	350	1.3	40,800	30,000	3,000	18,000
GE Oil & Gas Nuovo Pignone	MCL					x	x						OF		294,290 (8,333)	580	40	1.3	54,400	40,000	3,000	20,000
GE Oil & Gas Nuovo Pignone	DCL					x	x						OF		58,860 (1,667)	1,890	130	1.3	95,200	70,000	3,600	15,000
GE Oil & Gas Nuovo Pignone	SRL (overhung, single stage)				x								OF		206,000 (5,833)	1,015	70	1.3	20,400	40,000	3,000	13,000
GE Oil & Gas Nuovo Pignone	RA				x				x				OF/OI	230 (6)	58,860 (1,667)	1,015	70		20,400	15,000	1,500	50,000
GE Oil & Gas Thermodyn	BCL-LP/MP					x		x					OF	700 (20)	58,860 (1,667)	2,900	200		20,000	15,000	10,000	20,000
GE Oil & Gas Thermodyn	ICL					x	x						OF		29,430 (833)	2,075	200		20,000	15,000	1,500	14,000
Hitachi, Ltd./Hitachi Plant Technologies, Ltd.	2BCH					x		x					OF	88 (2)	56,500 (1,600)	10,900	750	1.3	67,000	50,000	1,200	14,000
Hitachi, Ltd./Hitachi Plant Technologies, Ltd.	2MCH					x	x						OF	88 (2)	204,800 (5,800)	650	45	1.3	67,000	50,000	3,000	18,000
Hitachi, Ltd./Hitachi Plant Technologies, Ltd.	3MCH					x	x						OF	280 (8)	204,800 (5,800)	650	45	1.3	67,000	50,000	3,500	18,000
Hitachi, Ltd./Hitachi Plant Technologies, Ltd.	BCH					x		x					OF	88 (2)	56,500 (1,600)	10,900	750	1.3	67,000	50,000	3,000	18,000
Hitachi, Ltd./Hitachi Plant Technologies, Ltd.	MCH					x	x						OF	280 (8)	204,800 (5,800)	650	45	1.3	67,000	50,000	2,500	18,000
Hitachi, Ltd./Hitachi Plant Technologies, Ltd.	POH					x	x						OF	1,770 (50)	35,300 (1,000)	650	45	1.3		25,000	3,000	14,000
Hitachi, Ltd./Hitachi Plant Technologies, Ltd.	POB-CH					x		x					OF	1,770 (50)	35,300 (1,000)	1,740	120	1.3	35,500	25,000	3,000	14,000
Hitachi, Ltd./Hitachi Plant Technologies, Ltd.	POB-GH					x		x					OF	1,770 (50)	35,300 (1,000)	1,160	80	1.3	35,500	20,000	3,000	14,000
Howden Process Compressor	SG26				x				x				OF	4,238 (120)	6,357 (180)	29	2	3	268	200	5,000	33,000
Howden Process Compressor	SG35				x				x				OF	4,238 (120)	8,476 (240)	29	2	3	603	450	5,000	33,000
Howden Process Compressor	SG40				x				x				OF	4,238 (120)	10,595 (300)	29	2	3	671	500	5,000	33,000
Howden Process Compressor	SG45				x				x				OF	8,476 (240)	19,070 (540)	29	2	3	1,341	1,000	5,000	33,000

Data Courtesy COMPRESSORTechTwo magazine. Additional current data available at www.compressortech2.com or www.CTSSNet.net

Dynamic Compressor Specifications

Manufacturer	Model Designation	OF = Oil Free / OI = Oil Injected	Inlet Flow min – acfm (m3/min)	Inlet Flow max – acfm (m3/min)	MAWP psig	MAWP bar	Compression Ratio (per stage)	Max Input HP	Max Input kW	speed_min RPM	speed_max RPM
Howden Process Compressor	SG52	OF	12,713 (360)	25,427 (720)	29	2	3	2,146	1,800	5,000	33,000
Howden Process Compressor	SG60	OF	14,832 (420)	31,784 (900)	29	2	3	2,414	1,800	5,000	33,000
Howden Process Compressor	SG65	OF	16,951 (480)	33,903 (960)	29	2	3	2,682	2,000	5,000	33,000
Howden Process Compressor	SG70	OF	21,189 (600)	40,259 (1,140)	29	2	3	3,487	2,600	5,000	33,000
Howden Process Compressor	SG80	OF	25,427 (720)	52,973 (1,500)	29	2	3	4,023	3,000	5,000	33,000
Howden Process Compressor	SG92	OF	33,903 (960)	63,567 (1,800)	29	2	3	5,364	4,000	5,000	33,000
Howden Process Compressor	SG105	OF	44,497 (1,260)	84,757 (2,400)	29	2	3	6,705	5,000	5,000	33,000
Ingersoll Rand	Centac CH	OF	800 (23)	6,000 (170)	35	3	3	600	448		3,600
Ingersoll Rand	Centac 2CIIDF	OF	3,300 (35)	9,000 (255)	35	3	3	900	671		3,600
Ingersoll Rand	Centac CV	OF	800 (23)	2,450 (69)	135	9	3	500	373		3,600
Ingersoll Rand	Centac EPF/EPC	OF	800 (28)	2,300 (65)	255	18	3	800	597		3,600
Ingersoll Rand	Centac C250/C350	OF	1,100 (31)	2,700 (76)	150	10	2	500	373		3,600
Ingersoll Rand	Centac C750	OF	1,800 (50)	2,100 (60)	610	43	3	1,000	746		3,600
Ingersoll Rand	Centac C700	OF	1,900 (55)	4,200 (122)	150	10	2	1,750	1,306		3,600
Ingersoll Rand	Centac 2CIISB	OF	3,000 (85)	4,600 (130)	350	24	3	1,500	671		3,600
Ingersoll Rand	Centac C1050	OF	4,200 (108)	3,800 (108)	610	43	3	1,500	1,119		3,600
Ingersoll Rand	Centac C950	OF	4,200 (122)	8,000 (170)	150	10	2	2,000	1,492		3,600
Ingersoll Rand	Centac 3CII	OF	6,000 (170)	9,000 (255)	175	12	2	2,000	2,811		1,800
Ingersoll Rand	Centac C3000	OF	9,000 (255)	15,000 (425)	150	10	2	3,500	4,476		1,800
Ingersoll Rand	Centac 5CII	OF	12,500 (354)	30,000 (849)	150	10	2	4,476	4,476		1,800
Ingersoll Rand	Centac 3C/4C	OF	5,000 (142)	15,000 (425)	375	26	3	4,500	3,357		1,800
Kobe Steel (Kobelco)	VH Series	OF	1,059 (30)	58,951 (1,670)	5,000	350		28,000	20,000	5,000	18,000
Kobe Steel (Kobelco)	VGS / VGSP Series	OF	1,765 (50)	105,900 (3,000)	1,150	80		21,000	15,000	5,000	40,000
Kobe Steel (Kobelco)	DH Series	OF	1,059 (30)	50,126 (1,420)	1,300	90		14,000	10,000	1,500	18,000
Kobe Steel (Kobelco)	V-VS-VSS Series	OF	1,059 (30)	20,151 (5,670)	1,700	50		21,000	15,000	1,500	18,000
Kobe Steel (Kobelco)	VG / VGP Series	OF	1,765 (50)	105,900 (3,000)	1,150	90		21,000	15,000	5,000	60,000
MAN TURBO AG (Oberhausen)	A, AV	OF	29,000 (835)	900,000 (25,450)	900	25	3		120,000		
MAN TURBO AG (Oberhausen)	AG, AK, AKF	OF	17,500 (835)	765,000 (21,660)	280	19			125,000		
MAN TURBO AG (Oberhausen)	AR	OF	17,500 (835)	765,000 (21,660)	280	19			165,000		
MAN TURBO AG (Oberhausen)	HOFIM ™	OF	115 (4)	21,000 (695)	4,350	300	7		30,000		
MAN TURBO AG (Oberhausen)	RB	OF	140 (4)	136,000 (3,800)	14,500	1,000			80,000		
MAN TURBO AG (Oberhausen)	TURBAIR ® (RC)	OF	5,300 (150)	115,000 (3,300)	29	2			5,400		
MAN TURBO AG (Oberhausen)	RG	OF	1,500 (42)	210,000 (5,840)	3,200	225			60,000		
MAN TURBO AG (Oberhausen)	RH	OF	450 (13)	388,000 (11,000)	1,160	80			80,000		
MAN TURBO AG (Oberhausen)	RIK, RIKT, RIO	OF	1,000 (30)	18,000 (500)	300	21			100,000		
MAN TURBO AG (Oberhausen)	MOPICO ® (RM)	OF		247,000 (7,000)	2,200	150		30,000	30,000		
MAN TURBO AG (Oberhausen)	Integrally Geared	OF		325,000 (9,200)	725	50	2.5	35,500	25,000		80,000
Mitsubishi	V Type	OF/OI	353 (10)	425,000 (12,000)	860	60		53,600	40,000	2,500	80,000
Mitsubishi	H Type	OF/OI	353 (10)	177,000 (5,000)	725	50	2	53,600	40,000	2,500	36,000
Mitsubishi	V Series	OI		35,300 (1,000)	9,400	650		42,900	32,000		
Mitsui Engineering & Shipbuilding	MA Series	OI		353,000 (9,200)	353	10	7	42,900	32,000		
Mitsui Engineering & Shipbuilding	(cont.)	OI	4 (1)	6,000 (170)	29	2		134,100	100,000		
Pedro Gil S.A.	RNT	OF			140		2	420	315		4,800
Rolls-Royce	RFBB-20	OF		12,720	2,000	140		20,000	14,900	9,000	13,800
Rolls-Royce	RFA-24	OF		25,300	2,000	140		20,000	14,900	9,000	13,800
Rolls-Royce	RFBB 30	OF		30,800	2,250	222		75,000	56,000	3,600	8,000
Rolls-Royce	RFBB-36	OF		45,400	2,250	155		75,000	56,000	3,600	6,686
Rolls-Royce	RFA 36	OF		60,500	1,800	125		75,000	56,000	3,800	6,500
Rolls-Royce	RFBB-42	OF		62,800	1,500	105		75,000	56,000	3,800	6,686
Rolls-Royce	RBB	OF		6,000	310			35,000	26,100	8,000	13,800
Rolls-Royce	RCB	OF		13,500	3,220	222		50,000	37,300	5,000	11,500
Rolls-Royce	RDB	OF		22,000	2,000	140		60,000	44,700	4,500	8,500
Rolls-Royce	REB	OF		35,500	2,000	85		75,000	56,000	3,500	6,500
SAMSUNG TECHWIN	SM Series	OF	12,713 (360)	27,016 (765)	363	25		7,200	5,369		1,800
SAMSUNG TECHWIN	SM7000	OF	8,723 (247)	12,496 (353)	363	25		3,200	2,386		3,600
SAMSUNG TECHWIN	SM6000	OF	4,838 (137)	8,193 (232)	363	25		2,130	1,588		3,600
SAMSUNG TECHWIN	SM5000	OF	3,426 (97)	4,838 (137)	363	25		1,280	955		3,600
SAMSUNG TECHWIN	SM4000	OF	1,836 (52)	3,178 (90)	363	25		850	634		3,600
SAMSUNG TECHWIN	SM3000	OF	953 (27)	1,766 (50)	125	8.5		350	261		3,600
SAMSUNG TECHWIN	SM2000	OF	494 (14)	812 (23)	125	8.5		175	130		3,600
SAMSUNG TECHWIN	SM1000	OF									
SAMSUNG TECHWIN	SD Series	OF									

Data Courtesy COMPRESSORTechTwo magazine. Additional current data available at www.compressortech2.com or www.CTSSNet.net

Dynamic Compressor Specifications

Manufacturer	Model Designation	Axial: Multi Stage	Axial: Fixed Stator Vanes	Axial: Variable Stator Vanes	Radial: Single Stage	Radial: Multiple Stage	Radial: Horizontally Split	Radial: Vertically Split	Radial: Integral Gear	Radial: Integral Electric	Thermal: Single Stage	Thermal: Multiple Stage	OF=Oil Free / IO=Oil Injected	Inlet Flow min — acfm (m3/min)	Inlet Flow max — acfm (m3/min)	MAWP psig	MAWP bar	Compression Ratio (per stage)	Max Input HP	Max Input kW	Speed min RPM	Speed max RPM
SAMSUNG TECHWIN	SD1000				x				x	x				459 (16)	706 (20)	100	7	7	150	112		59,000
SAMSUNG TECHWIN	SL Series																					
SAMSUNG TECHWIN	SL8000			x		x	x						OF	18,752 (531)	27,016 (765)	363	25		2,078	1,550		3,600
SAMSUNG TECHWIN	SL6000			x		x	x						OF	8,723 (247)	12,466 (353)	363	25		924	689		3,600
SAMSUNG TECHWIN	SL5000			x		x	x						OF	4,838 (137)	8,193 (232)	363	25		615	459		3,600
SAMSUNG TECHWIN	SL4000			x		x	x						OF	3,426 (97)	4,838 (137)	363	25		370	278		3,600
SAMSUNG TECHWIN	SL3000			x		x	x						OF	1,836 (52)	3,178 (90)	363	25		245	183		3,600
SAMSUNG TECHWIN	SL2000			x		x	x						OF	993 (27)	1,766 (50)	363	25		101	75		3,600
SAMSUNG TECHWIN	SL1000			x		x	x						OF	494 (14)	812 (23)	363	25		51	38		3,600
SAMSUNG TECHWIN	TM Series																					
SAMSUNG TECHWIN	TMX			x	x	x							OF	4,591 (130)	7,946 (225)	152	10.5		1,790	1,335		3,600
SAMSUNG TECHWIN	TMY		x		x	x							OF	3,249 (92)	4,662 (132)	363	25		930	694		3,600
SAMSUNG TECHWIN	TMZ		x		x	x							OF	1,695 (48)	3,037 (86)	363	25		610	455		3,600
SAMSUNG TECHWIN	SG Series	x																				
SAMSUNG TECHWIN	SG 3000	x			x								OF	854 (1,450)		95	7		250	186	1,800	3,600
Siemens AG	STC-GV				x				x				OF		8,000		200	2.8				45,000
Siemens AG	STC-GO			x	x				x				OF		7,196		60	1.8				43,000
Siemens AG	STC-SV					x		x					OF		8,000		1,000.00	1.7				20,000
Siemens AG	STC-SP					x		x					OF		2,000		150	1.4				15,000
Siemens AG	STC-SO					x		x					OF		10,000		60	1.5				61,000
Siemens AG	STC-SH					x		x					OF		10,000		50	1.7				20,000
Siemens AG	STC-GC				x				x				OF		2,000		60	2.2				40,000
Siemens AG	STC-Sx					x	x						OF		21,666		6	1.2				9,000
Siemens AG	STC-SR					x	x						OF		21,666		16	1.8				9,000
Siemens AG	STC-SI					x	x						OF		10,000		8	1.8				4,350
Siemens AG	STC-GVT				x				x				OF		8,000		8	2.6				45,000
Siemens AG	STC-STO				x								OF		2,166		2					30,000
Siemens AG	STC-GTO				x				x				OF		3,333		60	3.1				40,000
Siemens AG	STC-ECO				x													1.7		20,000		12,200
Solar Turbines Incorporated	C16				x				x					150 (5)	1,800 (50)	3,500	241	4	15,000	11,185		23,800
Solar Turbines Incorporated	C16 Series/Parallel				x				x					150 (5)	3,200 (90)	1,500	103	4	15,000	11,185		22,300
Solar Turbines Incorporated	C33				x				x					800 (25)	9,500 (270)	2,700	186		20,500	15,290		19,800
Solar Turbines Incorporated	C33 Series/Parallel				x				x					800 (25)	16,300 (460)	1,200	83		20,500	15,290		16,500
Solar Turbines Incorporated	C40 Multistage				x	x			x					800 (25)	9,000 (255)	2,500	172		20,500	15,290		14,300
Solar Turbines Incorporated	C41				x				x					150 (5)	15,000 (425)	3,750	259		20,500	15,290		12,000
Solar Turbines Incorporated	C50				x				x					2,200 (65)	20,000 (565)	1,500	103		20,500	15,290		14,000
Solar Turbines Incorporated	C51				x				x					1,000 (30)	15,000 (425)	3,000	207		30,000	22,370		12,000
Solar Turbines Incorporated	C61				x				x					5,000 (142)	15,000 (595)	2,700	207		30,000	22,370		10,170
Solar Turbines Incorporated	C40 Pipeline				x				x					1,200 (35)	9,500 (270)	1,800	186		10,310	7,690		10,500
Solar Turbines Incorporated	C45 Pipeline				x				x					2,800 (80)	13,000 (370)	1,600	124		20,500	15,290		12,000
Solar Turbines Incorporated	C85 Pipeline				x				x					3,500 (100)	20,000 (565)	1,600	110		20,500	15,290		10,500
Solar Turbines Incorporated	C85 Pipeline				x				x					8,000 (230)	45,000 (1,275)	1,600	110		30,000	22,370		7,000
Sundyne Corporation	LMC				x				x				OF	50 (85)	8,000 (13,600)	2,160	149	4	400	300	2,950	34,000
Sundyne Corporation	BMC				x				x				OF	50 (85)	8,000 (13,600)	2,160	149	4	400	300	2,950	34,000
Sundyne Corporation	Pinnacle				x				x				OF	100 (170)	12,000 (20,400)	4,400	304	4	4,000	3,000	5,000	50,000
York/Frick (JCI)	M_25B				x				x				OF	400 (11)	1,800 (51)	600	41		2,500	1,865		24,990
York/Frick (JCI)	M_25A				x				x				OF	1,000 (28)	2,650 (75)	600	41		2,500	1,865		20,610
York/Frick (JCI)	M_26B				x				x				OF	2,100 (60)	3,900 (110)	600	41		5,000	3,730		16,900
York/Frick (JCI)	M_26A				x				x				OF	3,300 (94)	5,600 (159)	600	41		5,000	3,730		13,930
York/Frick (JCI)	M_38B				x				x				OF	4,600 (130)	8,400 (238)	600	41		10,800	8,055		11,450
York/Frick (JCI)	M_38A				x				x				OF	6,700 (189)	12,200 (346)	600	41		10,800	8,055		9,410
York/Frick (JCI)	M_55B				x				x				OF	10,000 (283)	18,500 (524)	450	31		17,180	12,900		7,720
York/Frick (JCI)	M_55A				x				x				OF	14,200 (402)	23,000 (651)	450	31		17,300	12,900		6,540

Data Courtesy COMPRESSORTechTwo magazine. Additional current data available at www.compressortech2.com or www.CTSSNet.net

Positive Displacement Compressor Specifications

Manufacturer	page	Model Designation	Recip. Type	OF/OI	Inlet Flow min acfm(m3/min)	Inlet Flow max acfm(m3/min)	MAWP psig	MAWP bar	Max Rod Load-lb	Max Rod Load Newtons	Compression Ratio per stage	Max Input Power HP	Max Input Power kW	Speed min	Speed max
Ariel Corporation	1	JGM:P	Recip/Separable/Bal.Opposed	OI			6,100	421	7,000	31,138	4	170	127		1,800
Ariel Corporation		JGN:Q		OI			6,100	421	11,000	48,930		280	209		1,800
Ariel Corporation		JG:A		OI			6,100	421	11,000	48,930		840	626		1,800
Ariel Corporation		JGR:J		OI			6,100	421	23,000	102,309		1,860	1,387		1,800
Ariel Corporation		JGHE		OI			7,800	538	32,000	142,343		3,210	2,394		1,500
Ariel Corporation		JGK:T		OI			7,800	538	40,000	177,929		3,900	2,908		1,500
Ariel Corporation		JGC:D		OI			6,700	462	60,000	266,893		6,210	4,631		1,200
Ariel Corporation		JGU:Z		OI			6,700	462	80,000	355,858		7,800	5,816		1,200
Ariel Corporation		KBB:V		OI			6,700	462	100,000	444,822		10,000	7,457		900
Ariel Corporation		RG282M	Helical Lobe Screw	OF		3,035.00 (88.00)	230	16				800	597		4,065
Ariel Corporation		RG357M	Helical Lobe Screw	OF	1,000.00	5,225.00 (148.00)	230	16				1,375	1,025		3,230
Ariel Corporation		VRC-2		OF			1,300					130			1,800
Arrow Engine Company		H		OF/OI	70.00 (2.00)	2,350.00 (67.00)	1,450	100			4	780	560	400	1,000
Blackmer		H0162/H0163		OF	7.1	16.9	335	23	2,650		5	10		400	825
Blackmer		H0172/H0173		OF	3.5	8.4	600	41	2,650		5	10		400	825
Blackmer		H0172/HDL173 (Watercooled)		OF	3.5	8.4	600	41	2,650		5	10		400	825
Blackmer		H0362/H0363		OF	15.3	36	335	23	3,400		5	15		400	825
Blackmer		H0362/HDL363 (Watercooled)		OF	15.3	36	335	23	3,400		5	15		400	825
Blackmer		H0322 (watercooled)		OF	3.8	9	1,000	69	3,400		5	15		400	825
Blackmer		HDL342/HDL343		OF	6.9	16.9	750	52	3,400		5	15		400	825
Blackmer		HDL372/HDL373		OF	10.2	24.1	600	41	3,400		5	40		400	825
Blackmer		HDL372/HDL373 (Watercooled)		OF	10.2	24.1	600	41	5,000		5	40		400	825
Blackmer		HD902/HD603		OF	10.2	24.1	335	23	5,000		5	40		400	825
Blackmer		HDL602/HDL603 (Watercooled)		OF	27.2	64.2	335	23	5,000		5	40		400	825
Blackmer		HDL642/HDL643		OF	27.2	64.2	750	52	5,000		5	40		400	825
Blackmer		HD612/HD613		OF	13.4	31.7	400	28	5,000		5	50		400	800
Blackmer		HDL612/HDL613 (Watercooled)		OF	22.9	53.7	400	28	7,000		5	50		400	800
Blackmer		HD942		OF	52.4	125.2	350	24	7,000		5	50		400	800
Blackmer		HDL942 (Watercooled)		OF	52.4	125.2	350	24	7,000		5	50		400	800
BORSIG ZM Compression GmbH	2	C200		OF/OI		1,130.00 (32.00)	4,350	300	7,000		4	335	250	600	1,500
BORSIG ZM Compression GmbH		BG090		OF/OI		1,695.00 (48.00)	8,702	600			4	200	149	585	1,500
BORSIG ZM Compression GmbH		BG110		OF/OI		2,895.00 (82.00)	8,702	600			4	295	220	473	1,200
BORSIG ZM Compression GmbH		BG140		OF/OI		8,760.00 (248.00)	8,702	600			4	2,350	1,755	382	900
BORSIG ZM Compression GmbH		BG180		OF/OI		20,130.00 (570.00)	8,702	600			4	3,538	301	237	750
BORSIG ZM Compression GmbH		BG220		OF/OI		30,760.00 (871.00)	8,702	600			4	3,595	2,540	250	600
BORSIG ZM Compression GmbH		BG300		OF/OI		45,200.00 (1,280.00)	8,702	600			4	5,695	4,250	200	469
BORSIG ZM Compression GmbH		BG360		OF/OI		68,450.00 (1,938.00)	8,702	600			4	14,445	10,780	150	375
BORSIG ZM Compression GmbH		BG450		OF/OI			14,500				4	13,000			333
Burckhardt Compression AG		Process Gas		OF			14,500					2,800			1,000
Burckhardt Compression AG		Labyrinth Piston		OF			50,000					47,600			257
Burckhardt Compression AG		Hyper		OF/OI											
Cameron's Compression Systems Group		AJax DPC 3404 LE		OI			5,500	379	40,000	54,233	6	970	722	265	440
Cameron's Compression Systems Group		AJax DPC 2804 STD		OI			5,500	379	40,000	54,233	6	845	630	265	440
Cameron's Compression Systems Group		AJax DPC 2804 LE		OI			5,500	379	40,000	54,233	6	800	597	265	440
Cameron's Compression Systems Group		AJax DPC 3403 LE		OI			5,500	379	40,000	54,233	6	726	541	265	440
Cameron's Compression Systems Group		AJax DPC 2803 STD		OI			5,500	379	40,000	54,233	6	633	472	265	440
Cameron's Compression Systems Group		AJax DPC 2803 LE		OI			5,500	379	40,000	44,742	6	600	447	265	440
Cameron's Compression Systems Group		AJax DPC 2802 STD		OI			5,500	379	33,000	44,742	6	465	347	265	440
Cameron's Compression Systems Group		AJax DPC 2802 LE		OI			5,500	379	33,000	44,742	6	422	315	265	440
Cameron's Compression Systems Group		AJax DPC 3401 LE		OI			5,500	379	30,000	40,675	6	384	298	265	440
Cameron's Compression Systems Group		AJax DPC 2801 STD		OI			5,500	379	30,000	40,675	6	232	173	265	440
Cameron's Compression Systems Group		AJax DPC 2801 LE		OI			5,500	379	30,000	40,675	6	192	143	265	440
Cameron's Compression Systems Group		AJax DPC 2202 STD		OI			5,500	379	30,000	40,675	6	295	220	265	440
Cameron's Compression Systems Group		AJax DPC 2202 LE		OI			5,500	379	30,000	40,675	6	295	220	265	440
Cameron's Compression Systems Group		AJax DPC 2201 STD		OI			5,500	379	30,000	40,675	6	147	110	265	440
Cameron's Compression Systems Group		AJax DPC 2201 LE		OI			5,500	379	30,000	40,675	6	147	110	265	440
Cameron's Compression Systems Group		SUPERIOR C-Force 34		OI			5,808	400	13,000	57,827		580	432	850	1,800
Cameron's Compression Systems Group		SUPERIOR C-Force 32		OI			5,808	400	13,000	57,827		290	216	850	1,800
Cameron's Compression Systems Group		SUPERIOR RAM52		OI			2,200	152	40,000	177,929		1,188	886	600	1,500
Cameron's Compression Systems Group		SUPERIOR RAM54		OI			2,200	152	40,000	177,929		2,375	1,772	600	1,800

Data Courtesy COMPRESSORTechTwo magazine. Additional current data available at www.compressortech2.com or www.CTSSNet.net

Positive Displacement Compressor Specifications

Manufacturer	page	Model Designation	Single Stage	Multiple Stage	Integral Engine Driven	Separable	Balanced Opposed	Sliding Vane	OF/OI	Max. Allow. Working Pressure (psig)	(bar)	Max Rod Load-lb	Max Rod Load Newtons	Comp. Ratio per stage	HP	kW	Speed min	Speed max
Cameron's Compression Systems Group		SUPERIOR MH62	X		X	X	X			8,000	552	52,000	231,308		1,800	1,343	600	1,200
Cameron's Compression Systems Group		SUPERIOR MH64	X		X	X	X			8,000	552	52,000	231,308		3,600	2,685	600	1,200
Cameron's Compression Systems Group		SUPERIOR MH66	X		X	X	X			8,000	552	52,000	231,308		5,400	4,027	600	1,200
Cameron's Compression Systems Group		SUPERIOR WH62	X		X	X	X			8,000	552	52,000	231,308		1,800	1,343	600	1,200
Cameron's Compression Systems Group		SUPERIOR WH64	X		X	X	X			8,000	552	65,000	289,134		3,600	2,685	600	1,200
Cameron's Compression Systems Group		SUPERIOR WH66	X		X	X	X			8,000	552	65,000	289,134		5,400	4,027	600	1,000
Cameron's Compression Systems Group		SUPERIOR WH72	X		X	X	X			8,000	552	65,000	289,134		1,700	1,268	600	1,000
Cameron's Compression Systems Group		SUPERIOR WH74	X		X	X	X			8,000	552	65,000	289,134		3,400	2,535	600	1,000
Cameron's Compression Systems Group		SUPERIOR WH76	X		X	X	X			8,000	552	65,000	289,134		5,100	3,803	600	1,000
Cameron's Compression Systems Group		SUPERIOR WG62	X		X	X	X			8,000	552	75,000	333,617		3,000	2,237	600	1,200
Cameron's Compression Systems Group		SUPERIOR WG64	X		X	X	X			8,000	552	75,000	333,617		6,000	4,474	600	1,200
Cameron's Compression Systems Group		SUPERIOR WG66	X		X	X	X			8,000	552	75,000	333,617		9,000	6,711	600	1,200
Cameron's Compression Systems Group	3	SUPERIOR WG72	X		X	X	X			8,000	552	75,000	333,617		2,500	1,864	600	1,000
Cameron's Compression Systems Group		SUPERIOR WG74	X		X	X	X			8,000	552	75,000	333,617		5,000	3,728	600	1,000
Cameron's Compression Systems Group		SUPERIOR WG76	X		X	X	X			8,000	552	75,000	333,617		7,500	5,593	600	1,000
Combined Heat & Power, Inc.		250						X	OI	300	21				30	22	900	2,400
Combined Heat & Power, Inc.		275						X	OI	300	21				40	30	900	2,400
Combined Heat & Power, Inc.		312						X	OI	300	21				40	30	900	2,400
Combined Heat & Power, Inc.		350						X	OI	1,100	76				150	112	900	2,400
Combined Heat & Power, Inc.		410						X	OI	1,100	76				225	168	900	2,400
Combined Heat & Power, Inc.		500						X	OI	1,100	76				338	252	900	2,400
Combined Heat & Power, Inc.		700						X	OI	1,100	76				900	671	900	1,800
Combined Heat & Power, Inc.		1000						X	OI	1,100	76				2,910	2,170	900	1,800
Combined Heat & Power, Inc.		1400						X	OI	1,200	83				4,500	3,356	600	900
Dresser-Rand		7ESHESV	X		X				OF/OI	12,000	827	11,500		52			70	225
Dresser-Rand		11TESH	X		X				OF/OI	12,000	827	25,000		112			134	225
Dresser-Rand		3MVR		X								150,000		672		3,000	2,237		225
Dresser-Rand		410KVR		X								150,000		672		3,750	2,797		225
Dresser-Rand		412KVSE		X								95,000		426		2,850	2,128		230
Dresser-Rand		412KVGSR		X								95,000		426		2,836	2,115		230
Dresser-Rand		512KVR		X								150,000		672		4,500	3,356		750
Dresser-Rand		AVIP	X			X			OF/OI	6,600	455	15,400	68,500	69		1,300	969		500
Dresser-Rand		BDC-OF8.5	X			X			OF/OI	12,000	827	38,500	8,656	172		4,000	2,983		600
Dresser-Rand		BDC-OF2H	X			X			OF/OI	12,000	827	90,000	20,234	403		8,200	6,115		600
Dresser-Rand		BDC-OF18H3	X			X			OF/OI	12,000	827	385,000	1,712,000	1,725		38,000	28,336		600
Dresser-Rand		BOS	X			X			OF/OI	6,600	455	90,000	400,320	108		11,250	8,389		500
Dresser-Rand		BVIP	X			X			OF/OI	6,600	455	24,200	107,640	148		2,125	1,585		600
Dresser-Rand		CVIP	X			X			OF/OI	6,600	455	33,000	146,780	161		2,880	2,148		600
Dresser-Rand		DVIP	X			X			OF/OI	8,800	607	60,000	206,160	202		5,000	3,729		500
Dresser-Rand		HHE-FB	X			X			OF/OI	12,000	827	36,000	160,150	108		5,625	4,195		
Dresser-Rand		HHE-VE	X			X			OF/OI	12,000	827	60,000	13,489	269		5,425	4,045		
Dresser-Rand		HHE-VG	X			X			OF/OI	12,000	827	72,000	16,187	323		10,630	7,927		
Dresser-Rand		HHE-VK	X			X			OF/OI	12,000	827	125,000	28,103	560		10,390	7,748		
Dresser-Rand		HHE-VL	X			X			OF/OI	12,000	827	140,000	31,475	627		22,500	####		
Dresser-Rand		HOS	X			X			OF/OI	12,000	827	220,000	49,460	986		7,200	5,369		500
Dresser-Rand		Super-HOS	X			X			OF/OI	6,600	455	60,000	266,880	269		7,200	5,369		500
Dresser-Rand		HSE	X			X			OF/OI	8,800	607	75,000	333,600	336		8,700	6,488		
Dresser-Rand		PHE	X			X			OF/OI	12,000	827	31,500	6,689	139		1,000	746		
Dresser-Rand	4	TCV-10	X			X				12,000	827	11,500	2,565	52		240	179		200
Dresser-Rand		TCV-12	X			X						130,000		582		4,700	3,505		200
Dresser-Rand		TCVD-8		X		X						130,000		582		5,600	4,177		200
Dresser-Rand		TCVD-10		X		X						130,000		582		4,080	3,043		200
Dresser-Rand		TCVD-12		X		X						130,000		582		5,100	3,804		200
Dresser-Rand		TLAD-6		X		X						100,000		448		6,125	4,568		200
Dresser-Rand		TLAD-8		X		X						100,000		448		2,400	1,790		200
Dresser-Rand		TLAD-10		X		X						100,000		448		3,200	2,387		200
Dresser-Rand		F		X					OF/OI	10,150	700				5	26,620	2,983		
Dresser-Rand		E		X					OF/OI	10,150	700				5	26,820	15,421		
Dresser-Rand		D		X					OF/OI	10,150	700				5	8,716	6,500		
Dresser-Rand		B		X					OF/OI	10,150	700								

Data Courtesy COMPRESSORTechTwo magazine. Additional current data available at www.compressortech2.com or www.CTSSNet.net

Positive Displacement Compressor Specifications

Manufacturer	page	Model Designation	Type(s) marked	OF/OI	Inlet Flow min acfm (m³/min)	Inlet Flow max acfm (m³/min)	MAWP psig	MAWP bar	Max Rod Load‑lb	Max Rod Load N	Compression Ratio per stage	HP	kW	Speed min	Speed max
Dresser‑Rand		A	Single/Multiple Stage, Balanced Opposed	OF/OI			10,150	700				5	4,693	3,500	
Dresser‑Rand		M	Single/Multiple Stage, Balanced Opposed	OF/OI			10,150	700					2,413	1,800	
Emerson Process		SZO22C3A	Sliding Vane	OI	13.5	27	215				13	16	32	2,400	4,800
Emerson Process		SZO4HC1A	Sliding Vane	OI	27	54	215				13	32		2,400	4,800
Emerson Process		SZO4HC1A	Sliding Vane	OI	27	54	400				13	32		2,400	4,800
Emerson Process		SZO56C1A	Sliding Vane	OI	35	70	400				13	32		2,400	4,801
Garo		AM, ASM, AB	Liquid Ring	OF	565.00 (16.00)	2,931.00 (83.00)	500				7	805	600	500	3,600
GE Oil & Gas AC		From 85S15 to 250xxL5		OF	600.00 (17.00)	40,000.00 (1,130.00)	600				9	13,400	10,000	1,300	15,000
GE Oil & Gas High Speed Reciprocating Compressors		A	Single/Multiple Stage, Separable, Balanced Opposed	OI	30.00 (0.85)	5,650.00 (160.00)	6,000	414	14,500	64,520	4	800	600	600	1,800
GE Oil & Gas High Speed Reciprocating Compressors		DS		OI	45.00 (1.30)	14,840.00 (420.00)	6,600	455	35,000	155,750	4	2,400	1,800	700	1,500
GE Oil & Gas High Speed Reciprocating Compressors		ES		OI	38.00 (1.10)	21,900.00 (620.00)	6,600	455	50,000	222,490	4	5,400		700	1,500
GE Oil & Gas High Speed Reciprocating Compressors		FS		OI	38.00 (1.10)	21,900.00 (620.00)	6,600	455	60,000	266,800	4	7,200	5,400	700	1,500
GE Oil & Gas High Speed Reciprocating Compressors		H		OI	27.50 (0.78)	1,943.00 (55.00)	6,000	414	10,000	44,480	4	300	300	900	1,800
GE Oil & Gas High Speed Reciprocating Compressors		M		OI	27.50 (0.78)	955.00 (27.00)	6,000	414	6,000	26,680	4	120	90	900	1,800
GE Oil & Gas Nuovo Pignone		AVTN		OF/OI	5.00 (0.14)	475.00 (13.00)	6,500	450	4,438	19,740	4	180	135	150	1,200
GE Oil & Gas Nuovo Pignone		BVTN		OF/OI	10.00 (0.28)	1,840.00 (50.00)	6,500	450	15,097	67,100	4	1,070	800	150	1,200
GE Oil & Gas Nuovo Pignone		HA		OF/OI	40.00 (1.10)	15,500.00 (425.00)	6,500	450	32,557	144,700	6	2,840	2,120	350	800
GE Oil & Gas Nuovo Pignone		HB		OF/OI	38.00 (1.00)	29,000.00 (800.00)	6,500	450	59,265	236,400	9	7,400	5,520	308	800
GE Oil & Gas Nuovo Pignone		HD		OF/OI	40.00 (1.10)	61,500.00 (1,700.00)	11,000	750	72,562	322,500	9	13,940	10,200	220	600
GE Oil & Gas Nuovo Pignone		HE		OF/OI	45.00 (1.20)	114,500.00 (3,100.00)	11,000	750	119,225	533,000	9	29,000	21,500	185	600
GE Oil & Gas Nuovo Pignone		HE‑S		OF/OI	45.00 (1.20)	122,300.00 (820.00)	11,000	750	150,750	670,000	9	22,000	18,200	300	600
GE Oil & Gas Nuovo Pignone		HF		OF/OI	66.00 (1.80)	143,500.00 (3,900.00)	11,000	750	195,435	868,600	9	38,250	28,500	144	470
GE Oil & Gas Nuovo Pignone		HG		OF/OI	97.00 (2.70)	144,000.00 (3,900.00)	11,000	750	348,750	1,550,000	9	55,030	41,000	116	450
GE Oil & Gas Nuovo Pignone		HM		OF/OI	57.00 (1.60)	16,500.00 (450.00)	6,500	450	39,717	176,520	6	4,830	3,600	280	800
GE Oil & Gas Nuovo Pignone		OA		OF/OI	20.00 (0.60)	3,150.00 (85.00)	6,500	450	26,325	117,000	6	580	435	300	650
GE Oil & Gas Nuovo Pignone		OC		OF/OI	20.00 (0.60)	4,250.00 (115.00)	6,500	450	37,935	168,600	6	940	700	250	650
GE Oil & Gas Nuovo Pignone	5	PH		OI	10.00 (0.28)	841.00 (23.00)	50,800	3,500	303,750	1,350,000	9	12,885	9,600	170	310
GE Oil & Gas Nuovo Pignone		PK		OI	12.00 (0.33)	1,201.00 (33.00)	50,800	3,500	562,500	2,500,000	9	80,538	60,000	170	310
GE Oil & Gas Nuovo Pignone		SHM	Helical Lobe Screw	OI	42.00 (1.20)	33,300.00 (920.00)	6,500	450	80,585	358,065	4	7,140		250	1,200
GE Oil & Gas Nuovo Pignone		SHMB	Helical Lobe Screw	OI	80.00 (2.30)	22,300.00 (610.00)	6,550	450	80,585	358,065	4	7,440	5,474	250	1,045
GE Oil & Gas Nuovo Pignone		CT	Helical Lobe Screw	OI	11.00 (2.00)	424.00 (12.00)	6,550	450	142,000	32,000	4	230	170	570	1,230
GreenField		CU		OF	32.00 (0.90)	194.00 (5.50)	7,250	500	80,000	18,000	4	120	90	580	1,385
GreenField		CC		OI	11.00 (0.30)	67.00 (1.90)	5,000	350	40,000	9,000	4	60	45	605	1,385
GreenField		CN		OI	9.00 (0.25)	34.00 (0.96)	5,000	350	33,300	7,500	4	27	20	735	1,200
GreenField		DM		OF	217.00 (6.00)	37.00 (1.05)	4,500	310	44,500	10,000	4	48	36	1,450	1,750
HAUG Kompressoren AG		oilless piston compressors — BMD		OF	10.50 (0.30)	11,654.00 (330.00)	17,600	500			4	5,364	4,000	700	1,500
Hitachi, Ltd‑Hitachi Plant Technologies, Ltd.		BSD		OF/OI	10.60 (0.30)	11,654.00 (330.00)	12,300	350			4	5,364	4,000	300	900
Hitachi, Ltd‑Hitachi Plant Technologies, Ltd.		BTO		OF/OI	10.50 (0.30)	11,654.00 (330.00)	35,300	1,000			4	5,364	4,000	300	900
Hitachi, Ltd‑Hitachi Plant Technologies, Ltd.		HSD		OF/OI	10.50 (0.30)	1,060.00 (30.00)	35,300				4	288	288	200	900
Howden BC Compressors		Burton Corblin » P series	Diaphragm	OF	3.50 (0.10)	11,750.00 (333.00)	3,600	250	49,500	220,000	4	3,400	2,500	200	1,050
Howden BC Compressors		Burton Corblin » D series	Diaphragm	OF	0.06 (0.00)	120.00 (3.50)	43,500	3,000	33,800	150,000	12	536	400	200	500
Howden BC Compressors		Burton Corblin » HPD series	Diaphragm	OF	35.00 (1.00)	412.00 (12.00)	6,500	450	23,600	105,000	8	940	700	200	500
Howden Process Compressor		xRV127/R1	Helical Lobe Screw	OI		207.00 (5.86)	300	20			10	200	150	1,800	3,500
Howden Process Compressor		xRV127/R3	Helical Lobe Screw	OI		280.00 (7.93)	300	20			10	200	150	1,800	3,600
Howden Process Compressor		xRV127/R5	Helical Lobe Screw	OI		345.00 (9.77)	300	20			10	200	150	1,800	3,600
Howden Process Compressor		xRV163/165	Helical Lobe Screw	OI		420.00 (11.89)	300	20			10	350	280	1,800	3,600
Howden Process Compressor		xRV163/193	Helical Lobe Screw	OI		500.00 (14.16)	300	20			10	350	280	1,800	3,600
Howden Process Compressor		xRV204/110	Helical Lobe Screw	OI		573.00 (16.23)	300	20			10	600	450	1,800	3,600
Howden Process Compressor		xRV204/145	Helical Lobe Screw	OI		756.00 (21.41)	300	20			10	600	450	1,800	3,600
Howden Process Compressor		xRV204/165	Helical Lobe Screw	OI		860.00 (24.36)	300	20			10	600	450	1,800	3,600
Howden Process Compressor		xRV204/193	Helical Lobe Screw	OI		952.00 (26.98)	300	20			10	600	450	1,800	3,600
Howden Process Compressor		GTV 228	Helical Lobe Screw	OI		682.00 (19.00)	870	60			3			1,250	3,600
Howden Process Compressor		WRV 163/145	Helical Lobe Screw	OI		397.00 (11.24)	350	24			12	470	350	1,500	4,500
Howden Process Compressor		WRV163/1.80	Helical Lobe Screw	OI		476.00 (13.54)	350	24			12	470	350	1,500	4,500
Howden Process Compressor		WRV204/1.10	Helical Lobe Screw	OI		570.00 (16.14)	350	24			12	766	766	1,500	4,500
Howden Process Compressor		WRV204/1.45	Helical Lobe Screw	OI		775.00 (21.95)	350	24			12	1,028	766	1,500	4,500
Howden Process Compressor		WRV204/1.65	Helical Lobe Screw	OI		855.00 (24.21)	350	24			12	1,028	766	1,500	4,500
Howden Process Compressor		WRV204/1.85	Helical Lobe Screw	OI		962.00 (27.24)	350	24			12	1,028	766	1,500	4,500
Howden Process Compressor		WRV 255/1.10	Helical Lobe Screw	OI		1,114.00 (31.55)	350	24			12	1,542	1,150	1,500	4,500
Howden Process Compressor		WRV 255/1.30	Helical Lobe Screw	OI		1,253.00 (35.48)	350	24			12	1,542	1,150	1,500	3,900

Data Courtesy COMPRESSORTechTwo magazine. Additional current data available at www.compressortech2. com or www.CTSSNet.net

Positive Displacement Compressor Specifications

Manufacturer	page	Model Designation	Recip. Single Stage	Recip. Multiple Stage	Helical Lobe Screw	OF/OI	Inlet Flow min acfm (m³/min)	Inlet Flow max acfm (m³/min)	Max Work. Press. psig	bar	Max Rod Load-lb	Comp. Ratio/stage	Max Input HP	Max Input kW	Speed min	Speed max
Howden Process Compressor		WRVi 255/1.45			x	OI		1,514.00 (42.88)	350	24		12	1,542	1,150	1,500	3,600
Howden Process Compressor	6	WRVi 255/1.85			x	OI		1,670.00 (47.29)	350	24		12	1,542	1,150	1,500	3,600
Howden Process Compressor		WRVi 255/1.93			x	OI		1,879.00 (53.21)	350	24		8	1,542	1,150	1,500	3,600
Howden Process Compressor		WRVi 255/2.20			x	OI		2,227.00 (63.07)	350	24		12	1,542	1,150	1,500	3,600
Howden Process Compressor		WRVi 321/1.32			x	OI		2,673.00 (75.70)	350	24		12	2,741	2,044	1,500	3,600
Howden Process Compressor		WRVi 321/1.85			x	OI		3,341.00 (94.62)	350	24		12	2,741	2,044	1,500	3,600
Howden Process Compressor		WRVi 321/1.93			x	OI		3,758.00 (106.43)	350	24		12	2,741	2,044	1,500	3,600
Howden Process Compressor		WRVi 321/2.20			x	OI		4,510.00 (127.72)	350	24		8	2,741	2,044	1,500	3,600
Howden Process Compressor		WRVi 365/1.45			x	OI		4,144.00 (117.36)	350	24		12	5,814	4,335	1,500	3,600
Howden Process Compressor		WRVi 365/1.85			x	OI		4,716.00 (133.56)	350	24		12	5,814	4,335	1,500	3,600
Howden Process Compressor		WRVi 365/1.93			x	OI		5,516.00 (156.21)	350	24		12	5,814	4,335	1,500	3,600
Howden Process Compressor		WRVi 385/1.93			x	OI		6,022.00 (170.54)	350	24		12	5,814	4,335	700	2,000
Howden Process Compressor		WRVT510/1.32			x	OI		7,528.00 (213.19)	350	24		12	6,700	5,000	700	2,000
Howden Process Compressor		WRVT510/1.65			x	OI		8,266.00 (234.09)	350	24		12	6,700	5,000	700	2,000
Howden Process Compressor		WRVT510/1.93			x	OI		8,266.00 (234.09)	350	24		4	6,700	5,000	7,500	15,000
Howden Process Compressor		H127/165			x	OF		794.00 (1,350.00)	288			4	288	214	7,500	15,000
Howden Process Compressor		HP204/110			x	OF		1,341.00 (2,273.00)	200	15		4	800	600	4,750	9,500
Howden Process Compressor		HD04/165			x	OF		2,012.00 (3,420.00)	130	10		4	800	600	4,750	9,500
Howden Process Compressor		HP255/110			x	OF		2,070.00 (3,516.00)	200	15		4	1,250	935	4,000	7,500
Howden Process Compressor		HD255/165			x	OF		3,106.00 (5,279.00)	130	10		4	1,250	935	4,000	7,500
Howden Process Compressor		HP408/110			x	OF		5,308.00 (2,597.00)	200	15		4	3,050	2,275	2,300	4,700
Howden Process Compressor		H408/165			x	OF		7,061.00 (13,532.00)	130	10		4	3,050	2,275	2,300	4,700
Howden Process Compressor		HP510/0/110			x	OF		8,266.00 (14,050.00)	200	15		4	4,000	2,985	2,000	3,750
Howden Process Compressor		H510/165			x	OF		12,397.00 (21,074.00)	130	10		4	4,000	2,985	2,000	3,750
Hycomp, Inc.		AN3A	x			OF	1.82	3.75	1,000		2,800	5	8		400	825
Hycomp, Inc.		AN4A	x			OF	2.84	5.86	1,000		2,800	5	8		400	825
Hycomp, Inc.		WN4A	x			OF	2.84	5.86	1,000		2,800	5	8		400	825
Hycomp, Inc.		AN6A	x			OF	4.09	8.43	1,000		2,800	5	8		400	825
Hycomp, Inc.		AN8	x			OF	4.09	8.43	1,000		2,800	5	8		400	825
Hycomp, Inc.		WN07	x			OF	0.45	0.94	1,500		2,800	5	8		400	825
Hycomp, Inc.		AN12	x			OF	8.18	16.87	200		2,800	5	11		400	825
Hycomp, Inc.		AN6C	x			OF	4.36	8.99	750		3,700	5	11		400	825
Hycomp, Inc.		WN6C	x			OF	4.36	8.99	750		3,700	5	11		400	825
Hycomp, Inc.		WN10C	x			OF	6.81	14.05	600		3,700	5	11		400	825
Hycomp, Inc.		AN10C	x			OF	6.81	14.05	600		3,700	5	11		400	825
Hycomp, Inc.		WN10C	x			OF	6.81	14.05	600		3,700	5	20		400	825
Hycomp, Inc.		AN25	x			OF	17.44	35.98	1,000		3,700	5	20		400	825
Hycomp, Inc.		WN26	x			OF	17.44	35.98	1,000		3,700	5	20		400	825
Hycomp, Inc.		AN12D	x			OF	7.95	18.4	2,000		4,700	5	20		400	825
Hycomp, Inc.		AN17D	x			OF	11.45	23.61	600		4,700	5	40		400	825
Hycomp, Inc.		AN23D	x			OF	15.56	32.14	600		4,700	5	40		400	825
Hycomp, Inc.	7	AH14E	x			OF	29.44	60.71	200		6,300	5	40		400	825
Hycomp, Inc.		WH14E	x			OF	9.09	18.74	800		6,300	5	40		400	825
Hycomp, Inc.		AN20E	x			OF	9.09	18.74	800		6,300	5	40		400	825
Hycomp, Inc.		WN20E	x			OF	13.08	26.98	600		6,300	5	40		400	825
Hycomp, Inc.		AN27E	x			OF	13.08	26.98	600		6,300	5	40		400	825
Hycomp, Inc.		WN27E	x			OF	17.81	36.73	600		6,300	5	40		400	825
Hycomp, Inc.		AN35E	x			OF	17.81	38.73	600		6,300	5	40		400	825
Hycomp, Inc.		AN44E	x			OF	23.26	47.97	600		6,300	5	40		400	825
Hycomp, Inc.		ANT2	x			OF	29.44	60.71	200		6,300	5	40		400	825
Hycomp, Inc.		WN90	x			OF	48.06	99.13	600		6,300	5	66		400	825
Hycomp, Inc.		WN20SF	x			OF	48.06	99.13	600		6,300	5	66		400	700
Hycomp, Inc.		AN20SF	x			OF	60.08	123.91	700		11,700	5	66		400	700
Hycomp, Inc.		AN27E	x			OF	19.19	33.59	700		11,700	5	66		400	700
Hycomp, Inc.		WN44F	x			OF	19.19	33.59	850		11,700	5	66		400	700
Hycomp, Inc.		AN44F	x			OF	29.07	50.88	650		11,700	5	66		400	700
Hycomp, Inc.		WN55F	x			OF	29.07	50.88	500		11,700	5	66		400	700
Hycomp, Inc.		AN75F	x			OF	36.8	64.39	400		11,700	5	66		400	700
Hycomp, Inc.		WN98	x			OF	50.08	87.85	200		11,700	5	66		400	700
Hycomp, Inc.		AH154	x			OF	60.08	105.14	300		11,700	5	66		400	700
Hycomp, Inc.		2ADIA		x		OF	102.21	178.87	250		11,700	5	66		400	700
Hycomp, Inc.		2AN9B		x		OF	2.84	5.86	1,000		2,800	5	8		400	825
Hycomp, Inc.						OF	4.09	8.43	500		2,800	5	8		400	825

Data Courtesy COMPRESSORTechTwo magazine. Additional current data available at www.compressortech2.com or www.CTSSNet.net

Positive Displacement Compressor Specifications

Manufacturer	Model Designation	Config. (Recip./Rotary)	OF/OI	Inlet Flow min — acfm (m3/min)	Inlet Flow max — acfm (m3/min)	Max Allow. Working Press. (psig)	Max Allow. Working Press. (bar)	Max Rod Load-lb	Max Rod Load Newtons	Compression Ratio per stage	Max Input Power HP	Max Input Power kW	Speed Range min	Speed Range max
Hycomp, Inc.	2AN8	Recip. Single Stage	OF	5.56	11.48	500		2,800		5	8		400	825
Hycomp, Inc.	2AN3C	Recip. Single Stage	OF	2.18	4.45	1,500		2,800		5	11		400	825
Hycomp, Inc.	2AN5C	Recip. Single Stage	OF	3.41	7.03	1,500		3,700		5	11		400	825
Hycomp, Inc.	2AN10C	Recip. Single Stage	OF	5.76	11.88	500		3,700		5	11		400	825
Hycomp, Inc.	2WN10C	Recip. Single Stage	OF	6.68	13.77	500		3,700		5	11		400	825
Hycomp, Inc.	2AN17	Recip. Single Stage	OF	11.04	22.77	500		3,700		5	23		400	825
Hycomp, Inc.	2AN10D	Recip. Single Stage	OF	6.72	13.86	750		4,700		5	23		400	825
Hycomp, Inc.	2AN15D	Recip. Single Stage	OF	10.18	20.99	750		4,700		5	23		400	825
Hycomp, Inc.	2AN26	Recip. Single Stage	OF	17.53	36.15	500		4,700		5	23		400	825
Hycomp, Inc.	2AN35	Recip. Single Stage	OF	22.9	47.22	500		4,700		5	23		400	825
Hycomp, Inc.	2WN35	Recip. Single Stage	OF	22.9	47.22	500		4,700		5	23		400	825
Hycomp, Inc.	2WN40	Recip. Single Stage	OF	26.17	53.97	500		4,700		5	43		400	825
Hycomp, Inc.	2WN40	Recip. Single Stage	OF	26.17	53.97	500		4,700		5	43		400	825
Hycomp, Inc.	2WN13E	Recip. Single Stage	OF	8.9	18.36	1,000		6,300		5	43		400	825
Hycomp, Inc.	2AN17E	Recip. Single Stage	OF	11.63	23.99	1,000		6,300		5	43		400	825
Hycomp, Inc.	2WN17E	Recip. Single Stage	OF	11.63	23.99	1,000		6,300		5	43		400	825
Hycomp, Inc.	2AN22E	Recip. Single Stage	OF	14.72	30.36	750		6,300		5	43		400	825
Hycomp, Inc.	2WN22E	Recip. Single Stage	OF	14.72	30.36	750		6,300		5	43		400	825
Hycomp, Inc.	2AN61	Recip. Single Stage	OF	40.89	84.33	350		6,300		5	43		400	825
Hycomp, Inc.	2WN61	Recip. Single Stage	OF	40.89	84.33	350		6,300		5	72		400	700
Hycomp, Inc.	2AN76	Recip. Single Stage	OF	51.11	105.41	350		6,300		5	72		400	700
Hycomp, Inc.	2WN76	Recip. Single Stage	OF	51.11	105.41	350		6,300		5	72		400	700
Hycomp, Inc.	2AN22F	Recip. Single Stage	OF	14.54	25.44	1,000		11,700		5	72		400	700
Hycomp, Inc.	2WN22F	Recip. Single Stage	OF	14.54	25.44	1,000		11,700		5	72		400	700
Hycomp, Inc.	2WN28F	Recip. Single Stage	OF	18.4	32.2	1,000		11,700		5	72		400	700
Hycomp, Inc.	2AN28F	Recip. Single Stage	OF	18.4	32.2	750		11,700		5	72		400	700
Hycomp, Inc.	2WN38F	Recip. Single Stage	OF	25.04	43.82	750		11,700		5	72		400	700
Hycomp, Inc.	2WN49F	Recip. Single Stage	OF	32.71	57.24	750		11,700		5	72		400	700
Hycomp, Inc.	2AN58F	Recip. Single Stage	OF	38.39	67.18	750		11,700		5	72		400	700
Hycomp, Inc.	2AN137	Recip. Single Stage	OF	90.86	159	200		11,700		5	72		400	700
Hycomp, Inc.	2WN150L	Recip. Single Stage	OF	101.32	177.31	500		11,700		5	72		400	700
Hycomp, Inc.	2WN150H	Recip. Single Stage	OF	101.32	177.31	400		11,700		5	72		400	700
Hycomp, Inc.	3AN44	Recip. Single Stage	OF	29.44	80.71	500		4,700		5	23		400	825
J.P. Sauer & Sohn	WP150L	Recip.	OI	40.00 (1.10)	100.00 (2.80)	600	40			3	50	37	970	1,800
J.P. Sauer & Sohn	WP276L	Recip.	OI	75.00 (2.10)	150.00 (3.00)	600	40			3	75	55	970	1,800
J.P. Sauer & Sohn	WP316L	Recip.	OI	90.00 (2.60)	200.00 (5.50)	600	40			3	100	75	970	1,800
J.P. Sauer & Sohn	WP3215	Recip.	OI	3.00 (0.10)	10.00 (0.30)	6,000	350			3	12		970	1,800
J.P. Sauer & Sohn	WP4325	Recip.	OI	7.00 (0.20)	20.00 (0.50)	6,000	400			4	17	13	970	1,800
J.P. Sauer & Sohn	WP4331	Recip.	OI	13.00 (0.40)	30.00 (0.90)	6,000	400			4	35	26	970	1,800
J.P. Sauer & Sohn	WP4351	Recip.	OI	24.00 (0.70)	55.00 (1.60)	5,000	350			4	47	35	970	1,800
J.P. Sauer & Sohn	WP4205	Recip.	OI	40.00 (1.10)	75.00 (2.10)	5,000	350			4	70	52	970	1,800
J.P. Sauer & Sohn	WP4320	Recip.	OI	200.00 (5.50)	765.00 (21.70)	3,500	250			4	215	160	970	1,800
J.P. Sauer & Sohn	WP4440	Recip.	OI	90.00 (2.50)	470.00 (13.30)	6,000				4	215	160	970	1,800
J.P. Sauer & Sohn	WP0205	Recip.	OI	70.00 (2.00)	115.00 (3.30)	750	50			4	335	250	970	1,800
J.P. Sauer & Sohn	WP5310	Recip.	OI	320.00 (9.00)	880.00 (25.00)	500				4	270	200	970	1,800
J.P. Sauer & Sohn	WP5440	Recip.	OI	190.00 (5.40)	350.00 (10.00)	6,000	400			4	300	230	970	1,800
J.P. Sauer & Sohn	WP5550	Recip.	OI	115.00 (3.30)	210.00 (6.00)	7,250	500			4	175	132	970	1,800
Kobe Steel (Kobelco)	KR Series	Rotary/Recip.	OF/OI		65,000.00	10,150	700	220,000	980,000	10	27,000	20,000		1,000
Kobe Steel (Kobelco)	Dry Screw	Rotary	OF		12,000.00	650	45			15	13,500	10,000	1,500	15,000
Kobe Steel (Kobelco)	Oil-Injected Screw	Rotary	OI			8,700	100				10,000	7,500	1,000	10,000
Leobersdorfer Maschinenfabrik AG (LMF)	Process Gas	Recip./Rotary	OF/OI			8,700				3	8,000			1,200
Leobersdorfer Maschinenfabrik AG (LMF)	EcoPET		OF/OI			580				4	778			980
Leobersdorfer Maschinenfabrik AG (LMF)	Compound		OI			6,000				4	1,743			1,800
Leobersdorfer Maschinenfabrik AG (LMF)	CING		OI			5,075				4	800			1,800
Leobersdorfer Maschinenfabrik AG (LMF)	Air & Water Cooled Heavy Duty		OF/OI			7,250				4	800			1,800
LeROI-RCT	HG10000		OI			350	24				90	67	2,650	5,400
LeROI-RCT	HGF10000		OI			350	24				90	67	1,770	3,600
LeROI-RCT	HG12000		OI			350	24				125	93	2,250	5,200
LeROI-RCT	HG12xxx		OI			350	24				125	93	1,500	6,000
LeROI-RCT	HGF12000		OI			350	24				125	93	750	3,500

Data Courtesy COMPRESSORTechTwo magazine. Additional current data available at www.compressortech2.com or www.CTSSNet.net

Positive Displacement Compressor Specifications

Manufacturer	page	Model Designation	Recip. Single Stage	Recip. Multiple Stage	Recip. Integral Engine Driven	Recip. Separable	Recip. Balanced Opposed	Recip. Diaphragm	Rot. Straight Lobe	Rot. Helical Lobe Screw	Rot. Single	Rot. Sliding Vane	Rot. Scroll	Rot. Trochoidal	Rot. Liquid Ring	OF/OI	Inlet Flow min	Inlet Flow max	Press. psig	Press. bar	Max Rod Load-lb	Max Rod Load N	Compr. Ratio/stage	HP	kW	Speed min	Speed max
LeROI-RCT		HG17xxx	x							x						OI			300	21				175	130	540	4,000
LeROI-RCT		HGS17xxx	x							x						OI			350	24				175	130	540	3,560
LeROI-RCT		HGSH17000	x							x						OI			350	24				175	130	540	1,100
LeROI-RCT		HG24xxx	x							x						OI			250	17				350	261	500	3,600
LeROI-RCT		LG30xxx	x							x						OI			200	14				600	447	350	2,400
LeROI-RCT		LGL30xxx	x							x						OI			150	10				800	597	350	2,400
LeROI-RCT		HGT17		x						x						OI			500	34				400	298	750	3,600
LeROI-RCT		HGT24		x						x						OI			500	34				400	447	500	3,600
MAN TURBO AG (Oberhausen)		CP	x													OF	120.00 (3.00)	12,000.00 (340.00)	725	50			7	4,000	3,000	3,500	24,000
MAN TURBO AG (Oberhausen)		Skuel		x												OF	2,000.00 (58.00)	60,000.00 (1,700.00)	230	16			7	8,700	6,500	2,000	9,000
Mehrer Kompressoren GMBH & Co.		TE	x													OF			580	40			20	100	55	400	700
Mehrer Kompressoren GMBH & Co.		TZ		x												OF			580	21			30	72	75	90	600
Mehrer Kompressoren GMBH & Co.		TVB 700	x													OF			854	60			20	270	90	400	1,000
Mehrer Kompressoren GMBH & Co.		TVB 900	x													OF			854	60			8		200	400	1,000
Mehrer Kompressoren GMBH & Co.		TVE 700	x													OF			100	7			20	270	90	400	1,000
Mehrer Kompressoren GMBH & Co.		TVE 900	x													OF			100	7			20		200	400	1,000
Mehrer Kompressoren GMBH & Co.		TVZ 700		x												OF			363	25			20	270	90	400	1,000
Mehrer Kompressoren GMBH & Co.		TVZ 900		x												OF			363	25			20		200	400	1,000
Mehrer Kompressoren GMBH & Co.		TVD 700		x												OF			854	60			20		200	400	1,000
Mehrer Kompressoren GMBH & Co.		TVD 900		x												OF	185		854	60			15	270	200	400	1,000
Mehrer Kompressoren GMBH & Co.		BST (air)	x													OF		1,785.00	142	10			15	480	355	600	1,000
Mehrer Kompressoren GMBH & Co.		TRZ		x												OF			363	25							
Mehrer Kompressoren GMBH & Co.		TRE		x												OF			102	7							
Mehrer Kompressoren GMBH & Co.		TRB		x												OF			580	40							30,000
Mitsui Engineering & Shipbuilding		C Series	x													OF/OI	150.00 (4.00)	14,125.00 (400.00)	21,200	600			7	24,138	18,000	500	500
Mitsui Engineering & Shipbuilding		MB Series	x													OF/OI	14,125.00 (400.00)	14,125.00 (400.00)	21,200	600			7	134	1,000	700	700
MYCOM CANADA LTD.		V series	x							x						OI	150.00 (4.00)	8,100.00 (230.00)	350	25		890,000	20	2,500	1,900	1,500	6,000
MYCOM CANADA LTD.		C series	x							x						OI	250.00 (7.00)	2,300.00 (65.00)	350	25		890,000	30	2,000	1,500	1,500	6,000
MYCOM CANADA LTD.	10	GR series	x							x						OI	250.00 (7.00)	600.00 (17.00)	260	19			20	300	225	1,800	3,600
MYCOM CANADA LTD.		GH series	x							x						OI	250.00 (7.00)	2,300.00 (65.00)	750	50			20	2,700	1,100	1,800	3,600
MYCOM CANADA LTD.		VR series	x							x						OI	250.00 (8.00)	2,100.00 (60.00)	300	20			20	1,500	1,900	1,500	4,500
MYCOM CANADA LTD.		UD/G series	x							x						OI	150.00 (4.00)	8,100.00 (230.00)	350	25		712,000	20	2,500	220	1,500	6,000
MYCOM CANADA LTD.		FM series		x		x										OI	40.00 (1.00)	220.00 (12.00)	400	20		25,000	20	100	75	300	1,500
MYCOM CANADA LTD.		WA series		x		x										OI	220.00 (6.00)	670.00 (19.00)	300	20		80,000	20	300	225	600	1,500
MYCOM CANADA LTD.		WB series		x		x										OI	200.00 (6.00)	700.00 (20.00)	300	20		150,000	20	300	225	600	1,500
MYCOM CANADA LTD.		M series		x		x										OI	50.00 (1.00)	350.00 (11.00)	400	27		34,000	15	400	50	900	1,800
MYCOM CANADA LTD.		K series		x		x										OI	30.00 (1.00)	90.00 (3.00)	300	20		7,500	15	100	186	600	2,000
MYCOM CANADA LTD.		C series		x		x										OI	5.00 (0.14)	585.00 (16.60)	3,050	210		16,000	4	250	112	500	1,800
Natural Gas Services Group / SCS		PHT2		x	x											OF/OI	10.00 (0.28)	1,170.00 (33.20)	3,050	210	20,000	890,000	4	400	298	600	1,800
Natural Gas Services Group / SCS		PHT4		x	x											OF/OI	/3.00 (0.08)	270.00 (7.65)	2,000	138	20,000	890,000	4	400	298	500	1,800
Natural Gas Services Group / SCS		PVT2		x	x											OF/OI	6.00 (0.16)	540.00 (15.30)	2,000	138	7,500	34,000	4	100	75	600	2,000
Natural Gas Services Group / SCS		PVT4		x	x											OF/OI	5.00 (0.14)	585.00 (16.60)	3,050	210	7,500	34,000	4	250	112	500	1,800
Natural Gas Services Group / SCS		PxT2		x	x											OF/OI	10.00 (0.28)	1,170.00 (33.20)	3,050	210	16,000	712,000	4	250	186	650	1,800
Natural Gas Services Group / SCS		PxT4		x	x											OF/OI					16,000	712,000		135	186	600	1,800
NEUMAN & ESSER Deutschland GmbH		320 hts		x			x									OF/OI	105.00 (3.00)	105.00 (3.00)	710	50		25,000	4	135	100	200	700
NEUMAN & ESSER Deutschland GmbH		25	x				x									OF/OI	840.00 (24.00)	940.00 (24.00)	4,540	320		470,000	4	540	400	200	1,200
NEUMAN & ESSER Deutschland GmbH		30	x				x									OF/OI	3,500.00 (100.00)	3,500.00 (100.00)	9,800	700		80,000	4	540	400	200	1,200
NEUMAN & ESSER Deutschland GmbH		63	x				x									OF/OI	7,700.00 (220.00)	7,700.00 (220.00)	9,800	700		150,000	4	1,840	1,200	150	1,000
NEUMAN & ESSER Deutschland GmbH		130	x				x									OF/OI	13,000.00 (370.00)	13,000.00 (370.00)	9,800	700		380,000	4	3,400	2,500	150	1,200
NEUMAN & ESSER Deutschland GmbH		190	x				x									OF/OI	22,400.00 (640.00)	22,400.00 (640.00)	9,800	700		560,000	4	6,700	5,000	100	650
NEUMAN & ESSER Deutschland GmbH		300	x				x									OF/OI	37,500.00 (1,080.00)	37,900.00 (1,080.00)	9,800	700		800,000	4	9,800	7,300	300	600
NEUMAN & ESSER Deutschland GmbH		320 hts	x				x									OF/OI	150,000.00 (4,500.00)	150,000.00 (4,500.00)	9,800	700		1,500,000	4	40,000	30,000	200	600
NEUMAN & ESSER Deutschland GmbH		BV35	x				x									OF/OI	315.00 (9.00)	315.00 (9.00)	710	50		38,000	4	135	100	300	1,000
Pedro Gil S.A.		V1	x													OF/OI	1,120.00 (32.00)	1,120.00 (32.00)	3,550	250		110,000		250	250	200	600
Pedro Gil S.A.		RTN	x					x								OF	1.00 (0.03)	170.00 (48.00)	29	2				422	315	200	4,500
Quincy Compressor		QR-25 (3 to 25 HP)	x																									
Quincy Compressor		QT (3 to 15HP)	x																									
Quincy Compressor		QTS (1 to 5 HP)	x																									
Quincy Compressor		QRDS (5 to 30HP)	x																									
Quincy Compressor		QRDT (5 to 30 HP)		x																								

Positive Displacement Compressor Specifications

Manufacturer	page	Model Designation	Single Stage	Multiple Stage	Integral Engine Driven	Separable	Balanced Opposed	Diaphragm	Straight Lobe	Helical Lobe Screw	Single	Sliding Vane	Scroll	Trochoidal	Liquid Ring	OF = Oil Free/OI = Oil Injected	Inlet Flow min acfm (m³/min)	Inlet Flow max acfm (m³/min)	psig	bar	Max Rod Load-lb	Max Rod Load Newtons	Compression Ratio per stage	HP	kW	Speed min	Speed max
Quincy Compressor		QGS (5 to 15 HP)								x						OI	27.00 (0.77)	284.00 (8.04)	150	10			11	15	11	1,500	7,000
Quincy Compressor		QGB (7.5 to 60 HP)								x						OI	238.00 (6.74)	998.00 (28.27)	150	10			11	60	45	2,200	6,500
Quincy Compressor		QSF (50 to 200 HP)								x						OI	245.00 (6.94)	1,500.00 (42.49)	210	14			15	200	150	1,400	4,400
Quincy Compressor		QSI (50 to 300 HP)								x						OI	2,230.00 (63.17)	2,670.00 (75.63)	125	8			10	300	220		1,800
Quincy Compressor	11	QSI2 (400 to 500 HP)								x						OI								500	400		2,000
RO-FLO COMPRESSORS		4CC										x				OI	10.00 (0.25)	80.00 (2.30)	80	5			4	15	11	865	2,200
RO-FLO COMPRESSORS		5CC										x				OI	24.00 (0.68)	110.00 (3.18)	80	5			4	15	11	865	2,200
RO-FLO COMPRESSORS		7D										x				OI	53.00 (1.50)	198.00 (5.61)	80	5			4	33	25	690	1,465
RO-FLO COMPRESSORS		8D										x				OI	87.00 (2.46)	343.00 (9.71)	80	5			4	47	35	600	1,465
RO-FLO COMPRESSORS		8DE										x				OI	107.00 (3.03)	446.00 (12.63)	80	5			4	60	45	600	1,465
RO-FLO COMPRESSORS		10G										x				OI	126.00 (3.56)	674.00 (19.10)	80	5			4	103	77	450	1,300
RO-FLO COMPRESSORS		11S										x				OI	204.00 (5.78)	862.00 (24.41)	80	5			4	162	121	400	1,000
RO-FLO COMPRESSORS		11L										x				OI	232.00 (6.56)	986.00 (27.93)	80	5			4	168	125	400	1,000
RO-FLO COMPRESSORS		12S										x				OI	252.00 (7.13)	1,024.00 (29.01)	80	5			4	213	159	380	920
RO-FLO COMPRESSORS		12L										x				OI	280.00 (7.93)	1,159.00 (32.83)	80	5			4	222	166	380	920
RO-FLO COMPRESSORS		17S										x				OI	325.00 (9.21)	1,378.00 (39.05)	80	5			4	345	257	310	780
RO-FLO COMPRESSORS		17L										x				OI	393.00 (11.13)	1,594.00 (45.16)	80	5			4	400	298	310	780
RO-FLO COMPRESSORS		19S										x				OI	497.00 (14.08)	1,895.00 (53.70)	80	5			4	410	306	275	640
RO-FLO COMPRESSORS		19L										x				OI	560.00 (15.86)	2,165.00 (61.35)	80	5			4	390	291	275	640
RO-FLO COMPRESSORS		20B										x				OI	15.00 (0.43)	79.00 (2.23)	150	10			6	96	72	600	1,465
RO-FLO COMPRESSORS		208B										x				OI	19.00 (0.53)	140.00 (3.96)	150	10			6	340	254	400	1,300
RO-FLO COMPRESSORS		210M										x				OI	48.00 (1.36)	245.00 (6.95)	150	10			6	450	336	400	1,000
RO-FLO COMPRESSORS		211M										x				OI	58.00 (1.65)	278.00 (7.81)	150	10			6	450	336	380	920
RO-FLO COMPRESSORS		212M										x				OI	88.00 (2.50)	383.00 (10.85)	150	10			6	453	338	310	780
RO-FLO COMPRESSORS		217M										x				OI	97.00 (2.75)	474.00 (13.43)	150	10			6	478	356	275	640
SAFE S.R.L.		S3-S5-S7-S9	x	x		x										OF/OI						22,000	4			500	1,500
SAFE S.R.L.		SW		x		x	x									OI						50,000	4			550	1,500
SAFE S.R.L.		ST	x			x										OF						35,000	4			550	1,500
Sertco		350 LP	x		x	x										OF	50.00 (1.40)	900.00 (25.50)	350	10			6	120	90	900	2,100
Sertco		134 HP			x	x										OF	50.00 (1.40)	900.00 (25.50)	350	24			6	120	90	900	2,100
Sertco		100 HP			x	x										OF	50.00 (1.40)	900.00 (25.50)	350	34			6	120	90	900	2,100
SIAD Macchine Impianti S.p.A.		HD four		x		x	x									OI/OF		35,316.00 (1,000.00)	5,000	350			5	3,284	2,400		750
SIAD Macchine Impianti S.p.A.		HD two		x		x	x									OI/OF		17,658.00 (500.00)	5,000	350			5	1,652	1,200		750
SIAD Macchine Impianti S.p.A.		HPx four		x		x	x									OI/OF		21,186.00 (600.00)	5,000	350			5	1,340	1,000		750
SIAD Macchine Impianti S.p.A.		HPx two		x		x	x									OI/OF		10,595.00 (300.00)	5,000	350			5	804	600		750
SIAD Macchine Impianti S.p.A.		W series		x		x	x									OI/OF		1,766.00 (50.00)	3,610	250			5	335	250		1,200
SIAD Macchine Impianti S.p.A.		T series		x		x	x									OI/OF		1,235.00 (35.00)	3,610	250			5	102	75		1,200
SIAD Macchine Impianti S.p.A.		M series		x		x	x									OI/OF		3,550.00 (100.00)	3,610	250			5	603	450		1,200
SIAD Macchine Impianti S.p.A.		I series		x		x	x									OI/OF		530.00 (15.00)	2,900	200			5	14	10		1,200
Thomassen Compression Systems B.V.		C-7		x		x	x									OI/OF			8,700	600			5	1,090	800	300	600
Thomassen Compression Systems B.V.		C-12		x		x	x									OI/OF			8,700	600			5	3,130	2,300	300	600
Thomassen Compression Systems B.V.		C-25		x		x	x									OI/OF			8,700	600			5	7,500	5,600	300	600
Thomassen Compression Systems B.V.		C-35		x		x	x									OI/OF			8,700	600			5	14,000	10,300	250	500
Thomassen Compression Systems B.V.	12	C-45		x		x	x									OI/OF			8,700	600			5	20,950	15,400	250	500
Thomassen Compression Systems B.V.		C-85		x		x	x									OI/OF			8,700	600			5	33,460	24,000	190	375
Vilter Manufacturing, LLC		VSG 301									x					OI	310.00 (8.80)		515	35			20	400	298		4,800
Vilter Manufacturing, LLC		VSG 361									x					OI	356.00 (10.10)		515	35			20	400	298		4,800
Vilter Manufacturing, LLC		VSG 451									x					OI	408.00 (11.60)		515	35			20	400	298		4,800
Vilter Manufacturing, LLC		VSG 501									x					OI	495.00 (14.00)		515	35			20	600	448		4,800
Vilter Manufacturing, LLC		VSG 601									x					OI	590.00 (16.70)		515	35			20	600	448		4,800
Vilter Manufacturing, LLC		VSG 701									x					OI	680.00 (19.30)		515	35			20	600	448		4,800
Vilter Manufacturing, LLC		VSSG 291									x					OI	290.00 (8.20)		950	65			20	600	448		4,800
Vilter Manufacturing, LLC		VSSG 341									x					OI	341.00 (9.70)		950	65			20	600	448		4,800
Vilter Manufacturing, LLC		VSSG 451									x					OI	483.00 (13.70)		950	65			20	600	448		4,800
Vilter Manufacturing, LLC		VSSG 601									x					OI	568.00 (16.10)		950	65			20	865	645		4,800
Vilter Manufacturing, LLC		VSG 751									x					OI	789.00 (22.30)		485	33			20	865	645		4,800
Vilter Manufacturing, LLC		VSG 901									x					OI	892.00 (25.30)		485	33			20	865	645		4,800
Vilter Manufacturing, LLC		VSG 1051									x					OI	1,085.00 (30.70)		725	49			20	865	645		3,800
Vilter Manufacturing, LLC		VSG 1201									x					OI	1,210.00 (34.30)		725	49			20	1,400	1,044		3,800
Vilter Manufacturing, LLC		VSG 1551									x					OI	1,547.00 (43.80)		535	37			20	1,400	1,044		3,800
Vilter Manufacturing, LLC		VSG 1851									x					OI	1,815.00 (51.40)		535	37			20	1,400	1,044		3,800

Data Courtesy COMPRESSORTechTwo magazine. Additional current data available at www.compressortech2.com or www.CTSSNet.net

Positive Displacement Compressor Specifications

Manufacturer	page	Model Designation	Single Stage	Multiple Stage	Helical Lobe Screw	OF / OI	Inlet Flow Range max acfm (m3/min)	psig	bar	Compression Ratio per stage	HP	kW	Speed max
Vilter Manufacturing, LLC		VSG 2101				OF	2,048.00 (58.00)	535	37	20	1,400	1,044	3,800
VPT Kompressoren GmbH		RS	x			OI	318.00 (9.00)	190	13	13	63	47	3,600
VPT Kompressoren GmbH		WSV	x			OI	6,500.00 (110.00)	870	80	30	4,020	3,000	3,600
VPT Kompressoren GmbH		WST		x		OF	28,250.00 (800.00)	1,160	55	4	6,700	5,000	3,600
VPT Kompressoren GmbH		CD		x		OF	2,930.00 (83.00)	190	13	4	870	500	3,600
VPT Kompressoren GmbH		SKPT		x		OF	4,730.00 (134.00)	180	11	4	1,350	1,000	3,600
VPT Kompressoren GmbH		WSF	x			OI	2,120.00 (60.00)	380	26	26	1,350	1,000	3,550
York/Frick (JCI)		RWF-12			x	OI	72.00 (122.00)	334	23				5,772
York/Frick (JCI)		RWF-15			x	OI	89.00 (152.00)	334	23				4,661
York/Frick (JCI)		RWF-19			x	OI	111.00 (188.00)	334	23				3,550
York/Frick (JCI)		RWF-24			x	OI	144.00 (245.00)	334	23				5,772
York/Frick (JCI)		RWF-30			x	OI	180.00 (306.00)	334	23				4,661
York/Frick (JCI)		RWF-39			x	OI	223.00 (378.00)	334	23				3,550
York/Frick (JCI)		RWF-50			x	OI	292.00 (497.00)	334	23				6,297
York/Frick (JCI)		RWF-58			x	OI	341.00 (579.00)	334	23				4,306
York/Frick (JCI)		RWF-68			x	OI	403.00 (685.00)	334	23				4,306
York/Frick (JCI)		RWF-85			x	OI	499.00 (848.00)	334	23				3,600
York/Frick (JCI)	13	RWF-101			x	OI	596.00 (1,013.00)	334	23				4,500
York/Frick (JCI)		RWFII-100			x	OI	592.00 (1,005.00)	348	24				4,500
York/Frick (JCI)		RWFII-134			x	OI	790.00 (1,342.00)	348	24				4,500
York/Frick (JCI)		RWFII-177			x	OI	1,042.00 (1,770.00)	348	24				4,500
York/Frick (JCI)		RWFII-222			x	OI	1,311.00 (2,228.00)	348	24				4,500
York/Frick (JCI)		RWFII-270			x	OI	1,589.00 (2,700.00)	348	24				3,600
York/Frick (JCI)		RWFII-316			x	OI	1,865.00 (3,169.00)	348	24				4,500
York/Frick (JCI)		RWFII-399			x	OI	2,349.00 (3,992.00)	348	24				3,600
York/Frick (JCI)		RWFII-480			x	OI	2,624.00 (4,798.00)	348	24				3,600
York/Frick (JCI)		RWFII-496			x	OI	2,920.00 (4,961.00)	348	24				3,600
York/Frick (JCI)		RWFII-546			x	OI	3,216.00 (5,464.00)	348	24				3,600
York/Frick (JCI)		RWFII-676			x	OI	3,982.00 (6,765.00)	348	24				3,600
York/Frick (JCI)		RWFII-856			x	OI	5,058.00 (8,610.00)	348	24				3,600
York/Frick (JCI)		RWFII-1080			x	OI	6,394.00 (10,883.00)	348	24				3,600

Data Courtesy COMPRESSORTechTwo magazine. Additional current data available at www.compressortech2.com or www.CTSSNet.net

Appendix A2

COMPARISON TABLE OF THE THREE TYPES OF COMPRESSORS*

This chart provides a brief comparison of three compressor types: Reciprocating, Screw and Centrifugal relative to pressure and flow rate. For specific details per manufacturer and models the reader is directed to Appendix A.

Compressor Types		Reciprocating		Screw		Centrifugal
		Lube	Non-Lube	Oil Flooded	Oil Free	
Pressure	Maximum Discharge Pressure	4500 psiG (300barG)	1500 psiG (100 barG)	1500 psiG (100 barG)	600 psiG (40 barG)	3000 psiG (200barG)
	Maximum Pressure Ratio by Single Stage	3:1	3:1	> 50:1	4:1 - 7:1	1.5:1 - 3:1
Flow Rate	Maximum Actual Inlet Volume	8800ACFM (15000 m3/h)	8800ACFM (15000 m3/h)	15000ACFM (25000 m3/h)	41000ACFM (70000 m3/h)	240000ACFM (400000 m3/h)
	Turndown Accomplished by:	Suction Valve Unloaders (step & stepless); Clearance Pockets; Bypass	Suction Valve Unloaders (step & stepless); Clearance Pockets; Bypass	Slide Valve 15 - 100% (step & stepless); Bypass	(None); Bypass	Inlet Guide Vane; Speed Control (70 - 100%); Bypass
	Polymer Gas	Difficult	Difficult	Difficult	Possible	Difficult
	Dirty Gas	Possible	Difficult	Possible	Possible	Difficult
	MW Change	Possible	Possible	Possible	Possible	Difficult

*Process Gas Applications Where API 619 Screw Compressors Replaced Reciprocating And Centrifugal Compressors, Takao Ohama, Yoshinori Kurioka, Hironao Tanaka, Takao Koga; Proceedings of the Thirty-Fifth Turbomachinery Symposium—2006

Appendix B1

List of Symbols

A   Area, Ft^2

Btu   British thermal unit

C   Velocity of the air entering the compressor or air and combustion products leaving the turbine, *ft per sec*

CDP   Compressor discharge pressure, *psia*

CDT   Compressor discharge temperature, $°R$

c_p Specific heat at constant pressure, *Btu/lb°F*

c_p Specific heat at constant pressure, *Btu/lb°F*

c_v Specific heat at constant volume, *Btu/lb°F*

$°E$ Engler, degrees

g_c Acceleration due to gravity, *ft/sec²*

h_1 Enthalpy of the fluid entering, Btu/lb_m

h_2 Enthalpy of the fluid leaving, Btu/lb_m

h_i Enthalpy of the fluid entering, *Btu/lb*

h_o Enthalpy of the fluid leaving, Btu/lb_m

HP   Horsepower

Hr   Hours

J   Ratio of work unit to heat unit,   778.2 *ft* lb_f */Btu*

k   Ratio of specific heats, c_p/c_v

KE_1 kinetic energy of the fluid entering, Btu/lb_m

KE_2 Kinetic energy of the fluid leaving, Btu/lb_m

KJ   Kilojoules

KW   Kilowatts

Mo   Mass of a given volume of oil

Mw   Mass of a given volume of water

MW   Power output (electric generator)

MW   Molecular weight

N_1 Single spool or low pressure compressor-turbine rotor speed, rpm

N_2 Dual spool high pressure compressor-turbine rotor speed, rpm

N_3 Power turbine rotor speed, rpm

NO_x Oxides of nitrogen

P_2 Compressor inlet pressure, *psia*

P_3 Single spool compressor discharge pressure, single spool low pressure compressor discharge pressure, *psia*

P_4 Dual spool low pressure compressor discharge pressure, *psia*

P_5 Turbine inlet pressure, *psia*

P_3/P_2 Single spool compressor pressure ratio, dual spool low pressure compressor pressure ratio

P_4/P_2 Compressor pressure ratio

P_4/P_3 Dual spool high pressure compressor pressure ratio

P_{AMB} Atmospheric total pressure, *psia*

PE_1 Potential energy of the fluid entering, Btu/lb_m

PE_2 Potential energy of the fluid leaving, Btu/lb_m

P_i Inlet pressure, *psia*

P_o Discharge pressure, *psia*

cP   Absolute or Dynamic Viscosity, Centipoises

P_{PT} Power input to the load device, Btu/sec

Ps_3 Low pressure compressor out pressure (static)

Ps_4 High pressure compressor out pressure (STATIC)

P_{t7} Exhaust total pressure, *psia*

Q   Cubic feet per second (CFS)

$_1Q_2$ Heat transferred to or from the system, Btu/lb_m

Q_m Mechanical bearing losses of the driven equipment, Btu/sec

Q_m Mechanical bearing losses of the power turbine, Btu/sec

Q_r Radiation and convection heat loss , Btu/sec

R_c Compressor pressure ratio

secs R.I.   Redwood No 1 (UK), seconds

cST   Kinematic viscosity = Absolute viscosity/density. centistokes

SUS   Saybolt Universal (USA), seconds

T_{1DB} Dry bulb temperature upstream of the cooler, °R

T_{2DB} Dry bulb temperature downstream of the cooler, °R

T_{2WB} Wet bulb temperature downstream of the cooler, °R

T_2 Compressor inlet temperature, °R

T_3 Single spool compressor discharge temperature, dual spool low pressure compressor discharge temperature, °R

T_4 Dual spool low pressure compressor discharge temperature, °R

T_{amb} Ambient air temperature, °R

Ti Inlet temperature, °R

To Discharge temperature, °R

U   Velocity, *feet per sec*

D   Diameter, *in², ft²*

V   Vibration

W   Mass flow rate, *lb$_m$/sec*

W_2 Relative velocity

W_a Air flow entering the compressor, *lb/sec*

W_f Fuel flow, *lbs/sec*

W_f/CDP   Gas turbine burner ratio units, *lbs/hr-psia*

W_g Turbine inlet gas flow, *lbs/sec*

$_1W_2$ Work per unit mass on or by the system, *ft-lb$_f$/lb$_m$*

$_1W_2/J$ Work, *Btu/lb$_m$*

Z_{ave} Average compressibility factor of air,

β_2 Relative direction of air leaving the stator vanes

η_c Compressor efficiency

η_t Gas generator efficiency

η_{PT} Power extraction turbine efficiency

θ_1 Direction of air leaving the stator vanes

Appendix B2
Glossary of Terms

Absolute pressure—The total pressure measured from absolute zero in units such as pounds per square in absolute (psia), Inches of Mercury Absolute (in. of HGA).

Absolute temperature—The temperature of a body at absolute zero (Fahrenheit scale is minus 459.67°F/Celsius scale is minus 273.15°C).

Absolute viscosity (*Dynamic*)—Is the tangential force per unit area necessary to move one layer with respect to the other at unit velocity when maintained a unit distance apart by the fluid. Units are pound force sec per foot squared (lbf sec/ft$^{2)}$ or kilogram/meter sec (kg/m sec)

Absolute zero—The lower limit of temperature -459.69°F or -273.15°C

Acidity—The quality, state or degree of being acid. A substance is considered acidic if its pH is less than 7. In oils, acidity denotes the presence of acid-type constituents whose concentration is usually defined in terms of a neutralization number.

ACFM—Actual cubic feet per minute. The quantity of gas actually compressed and delivered at the temperature & pressure at that location.

Activated alumina—An adsorption type desiccant.

Adiabatic compression—Compression where no heat is transferred to or from the gas during the compression process.

Adiabatic efficiency—The ratio of the actual work output to the work output that would be achieved if the process between the inlet state and the exit state was isentropic.

Adsorptive filtration—The attraction to a filter medium by electrostatic forces, or by molecular attraction between the particles and the medium.

Aeration—A process by which air is circulated through or dissolved in a liquid or substance.

Aerosol—A suspension of fine solid or liquid particles in a gas.

Aftercooler—Heat exchangers for cooling air or gas discharged, typically, from compressors.

Air actuator—A device to convert pneumatic energy to mechanical energy.

Air bearing—A fluid dynamic (hydrostatic) bearing using air to support the rotating shaft loads within the bearing.

Air borne—Supported or transported by air.

Air cooled compressor—A compressor cooled by atmospheric air circulated around the cylinders or casing usually incorporating cooling fins.

Air flow—The movement of air from one point to another.

Air lock—An interlocking device that permits passage of air (or gas) between regions of different pressures whole maintaining the pressure in each region.

Air nozzle—A fitting incorporating an orifice whereby a liquid or gas passing through the orifice undergoes a change in pressure and velocity.

Air receiver—A receptacle that stores compressed air for heavy demands in excess of compressor capacity.

ALT—Altitude

Altitude—The elevation above sea level.

Alternating current (AC)—An electrical current that periodically reverses its direction. Standard in US and Canada is 60 cycles per second. Europe and other countries is 50 cycles per second.

Amagat's law—States that the volume of a mixture of gases is equal to the sum of the partial volumes which the constituent gases would occupy if each existed alone at the total pressure of the mixture.

Ambient—The temperature, pressure and humidity at specific site locations.

Ammonia—A colorless gas with a strong pungent odor made up of nitrogen and hydrogen (NH_3).

Amonton's law—States that the pressure of a gas, at constant volume, varies directly with the absolute temperature.

Ampere (AMP)—A unit of electrical current or rate of flow of electrons through a conductor. One volt across one ohm of resistance causes a current flow of one ampere.

Ancillary equipment—Components in addition to the primary equipment (such as the compressor, generator or driver).

Anhydrous—A substance containing no water.

Anion—An ion with more electrons than protons, giving it a net negative charge..

A.P.I.—American Petroleum Institute

Approach temperature—The difference in temperature between that of the heated medium exiting from a heat exchanger and the inlet temperature of the cooling medium entering the heat exchanger.

ASHRAE—The American Society of Heating, Refrigerating and Air-Conditioning Engineers, Inc. .

A.S.M.E.—American Society of Mechanical Engineers

A.S.T.M.—American Society For Testing and Materials.

Atmosphere (ATM)—The layer of gas surrounding the earth that is retained by gravity.

Atmospheres absolute (ATA)—It is the weight of the column of air existing above the earths surface at sea level. It is equivalent to 14.696 psia or 1.0333 kg/sq cm.

Atmospheric dew point—The temperature at which water vapor begins to condense at atmospheric pressure.

Attenuation—The gradual loss in intensity of sound waves due to distance or an blocking media (such as a wall).

Avogadro's Law—States that equal volumes of ideal gases, at the same temperature and pressure, contain the same number of molecules.

Axial compressor—A compressor belonging to the group of dynamic compressors. Characterized by having its flow in the axial direction.

Boyle's Law—States that for a fixed amount of an ideal gas kept at a fixed temperature, pressure and volume are inversely proportional.

Bag house—A dust-collection chamber containing numerous permeable fabric filters through which the gases pass.

Bar—A unit of pressure equal to 0.986925 atmospheres or 14.5039 psi.

Barometric pressure—Is the absolute atmospheric pressure existing at any given point in the atmosphere. It is the weight of a unit column of gas directly above the point of measurement. It varies with altitude, moisture and weather conditions.

Base plate—A metallic structure on which a compressor, motor, driver or other machine units are mounted.

Bernoulli's principle—States that for a fluid, with little or no viscosity, increases in the speed of the fluid flow occurs simultaneously with a decrease in pressure or a decrease in the fluids potential energy

BHP—Break horse power

Blower—A compressor designed to operate at lower pressure ratios.

Blow off control—When the maximum pressure is reached, the delivered air is blown off to the atmosphere.

Body—The stationary valve seating surface

Bonnet—The portion of a valve that surrounds the spring.

Booster compressor—Compressors used to increase the initial gas pressure, usually low or near atmospheric pressure, to a higher pressure.

Boyle's law—States that the volume of a gas, at constant temperature, varies inversely with the pressure.

Breather—A filtering unit for vented enclosures installed to prevent dirt and foreign matter from entering the enclosure.

British thermal unit—Btu. The amount of heat required to raise the temperature of one pound of water one degree Fahrenheit under set conditions of temperature and pressure.

Buna N—A synthetic rubber frequently used for vessel and liquid filter element gasket.

Burst pressure—Maximum pressure a vessel will withstand without bursting.

By pass—Condition that exists when the air, gas, or fluid normally passing through an element is being shunted around the element.

Capacity—Capacity of a compressor is the full rated volume of flow of gas compressed and delivered at certain set conditions.

Capacity gauge—A gauge that measures air flow as a percentage of capacity, used in rotary screw compressors as an estimator during modulation controls.

Carbon dioxide—A heavy colorless gas that does not support combustion but is formed by the combustion and decomposition of organic substances.

Carbon monoxide—A colorless odorless very poisonous gas formed by the incomplete burning of carbon.

Casing—The pressure containing stationary element that encloses the rotor and associated internal components of the compressor.

Cation—An ion with more protons than electrons.

Celsius—The international temperature scale where water freezes at 0 (degrees) and boils at 100 (degrees). Also known as the centigrade scale.

Centrifugal compressor—A dynamic compressor where air or gas is compressed by the mechanical action of rotating impellers and stationary diaphragms imparting velocity and pressure to the air or gas.

CFM—Cubic feet per minute. An airflow measurement of volume.

Charles Law—States that the volume of a gas is directly proportional to its temperature.

Chatter—Abnormal, rapid reciprocating movement of the disc on the seat of a pressure relief valve.

Check valve—A valve that permits flow in one direction only.

Choke—This term is used for turbo compressors and represents the maximum flow condition. It is sometimes also referred to as stonewalling.

Clean pressure drop—The pressure loss across the filter element determined under steady state flow conditions using a clean test fluid across a clean filter element.

Clearance—The maximum cylinder volume on a working side of the piston, minus the piston displacement volume per stroke. It is usually expressed as a percentage of the displaced volume.

Clearance pocket—An adjustable volume that may be opened or closed to the clearance space for increasing or decreasing the clearance, usually temporarily or seasonally, to reduce the volumetric efficiency and flow of the compressor.

Collapse pressure—The minimum differential pressure that an element is designed to withstand without permanent deformation.

CNG—Compressed natural gas.

Coalescing filter—A filter unit that combines three principles to filter out oil aerosols: 1) *Direct interception*—A sieving action, 2) *Inertial impaction*—Collision with filter media fibers, 3) *Diffusion* -Particles travel in a spiral motion, presenting an effective frontal area thus capturing particles within the filter medium.

Cogeneration—The use of exhaust heat flow as the source of heat to generate steam, hot water, hot oil, etc. to be used in another process.

Cold start—Starting a compressor from a state of total shutdown. Usually done with "local" control at the compressor.

Compressed air—Air under pressure greater than that of the atmosphere.

Compressibility—A measure of the relative volume change of a gas as a response to a pressure change.

Compressibility factor Z—Is the ratio of the actual volume of the gas to the volume determined according to the ideal gas law.

Compression efficiency—Is the ratio of the theoretical work requirement to the actual work required to be performed on the gas for compression and delivery.

Compression ratio—The ratio of the absolute discharge pressure to the absolute inlet pressure.

Condensate—The liquid phase that separates from a vapor during condensation.

Condenser—A heat exchanger used to change a vapor into a liquid.

Conduction—The transfer of thermal energy between neighboring molecules in a substance due to a temperature gradient

Constant speed control—A control method whereby load is varied either by use of recycle valves, suction valve unloaders or variable volume pockets and energy input at constant driver speed.

Contaminant—Foreign matter carried in the air, gas or fluid to be filtered out. Includes air borne dirt, metallic particles produced by wear of moving parts of the air compressor, rust from metal pipelines.

Contaminant capacity—An indicator of relative service life of a filter.

Control valve—Any valve that controls flow.

Convection—A means of transferring heat through mass flow. Also the transfer of heat within a fluid by movements of particles within the fluid.

Coolant—Fluid cooling agent.

Cooling tower—A cooling water supply system. There are two different types—Open and closed loop systems.

Coriolis force—A force acting in a direction perpendicular to the rotation axis and to the velocity of the body in the rotating frame and is proportional to the objects speed in the rotating frame.

CPM—Cycles per minute—a unit of measure of the frequency of any vibration.

Cracking—To subject petroleum oil to heat in order to break it down into lighter products.

Critical pressure—The pressure required at the critical temperature to cause the gas to change state. It is the highest vapor pressure that the liquid can exert.

Critical speed—Rotative speeds at which rotating machinery passes through its the natural frequency.

Critical temperature—That temperature above which a gas will not liquefy regardless of any increase in pressure.

Crosshead assembly—The assembly connecting the crankcase and connecting rod to the cylinder head and piston rod for translating circular to linear motion.

Crosshead loading

The tensile or compressive loading on the crosshead assembly with compressive piston rod loading on the outward stroke and tensile piston rod loading on the inward stroke.

CSA—Canadian Standards Association

Cubic feet per minute (CFM)—An airflow measurement of volume.

Cycle time—Amount of time for a compressor to complete one cycle.

Cylinder—The piston chamber in a compressor or actuator.

Cyclone—A type of separator for removal of larger particles from a gas stream. Gas laden with particulates enters the cyclone and is directed to flow in a spiral causing the entrained particulates to fall out and collect at the bottom. The gas exits near the top of the cyclone.

Cyclone separator—A means of purifying an gas stream by using both gravitational and centrifugal forces.

Dalton's law—States that the total pressure exerted by a gaseous mixture is equal to the sum of the partial pressures of each individual component in the gas mixture.

Degrees Celsius (°C)—An absolute temperature scale. ((°F − 32) x 5/9).

Degrees Fahrenheit (°F)—An absolute temperature scale. ((°C x 9/5) + 32).

Degrees Kelvin (°K)—An absolute temperature scale. The kelvin unit of thermodynamic temperature, is the fraction 1/273,16 of the thermodynamic temperature of the triple point of water. The triple point of water is the equilibrium temperature (0.01 °C or 273,16 K) between pure ice, air free water and water vapor.

Degree Rankine (°R)—An absolute temperature scale. (°F + 459,67).

Degree of saturation—The ratio of humidity ratio of moist air to humidity ratio of saturated moist air.

Deliquescent—Melting or dissolving and becoming a liquid by absorbing moisture.

Delta P—Describes the pressure drop through a component and is the difference in pressure between two points.

Delta T—Describes the temperature difference between two points.

Demand—Flow of a gas under specific conditions required at a particular point.

Demulsibility—The resistance of a hydraulic fluid to emulsification.

Density—The weight of a given volume of gas, usually expressed in lb/ cu ft at standard pressure and temperature conditions.

Desiccant—An adsorption type material used in compressed air dryers. Industry standards are activated alumina, silica gel and molecular sieves.

Design pressure—The maximum continuous operating pressure as designed by the manufacturer.

Desorption—Opposite of absorption or adsorption.

Dew point—Of a gas is the temperature at which the vapor in a space (at a given pressure) will start to condense (form dew). Dew point of a gas mixture is the temperature at which the highest boiling point constituent will start to condense.

Diaphragm—A stationary element between rotating stages of a multistage centrifugal compressor.

Diaphragm compressor—Is a positive displacement reciprocating compressor using a flexible membrane or diaphragm in place of a piston.

Diaphragm cooling—A method of removing heat from the flowing medium by circulation of a coolant in passages built into the diaphragm. Also known as diaphragm routing.

Differential pressure—The difference in pressure between any two points of a system or component.

Differential pressure indicator—An indicator, or differential pressure gage, which signals the difference in pressure between any two points of a system or a component.

Diffuser—A stationary passage surrounding an impeller, in which velocity pressure imparted to the flow medium by the impeller is converted into static pressure.

Digital controller—Control device whose operation may be reduced to binary operation such as ON = 0, OFF = 1 or OPEN and CLOSED. Also known as Logic controls

Diluent—An inert substance added so as to reduce the activity of a substance.

Direct current—DC. A continuous, one directional flow of electricity.

Directional control valve—A valve to control the flow of a gas in a certain direction.

Dirt holding capacity—The quantity of contaminant a filter element can trap and hold before the maximum allowable back pressure or delta P level is reached.

Disc—The movable seating surface in a valve.

Discharge piping—The piping between the compressor and the aftercooler, the aftercooler separator and the downstream element.

Discharge pressure—The total gas pressure (static plus velocity) at the discharge port of the compressor. Velocity pressure is considered only with dynamic compressors.

Discharge temperature—The temperature existing at the discharge port of the compressor.

Displacement compressor—A machine where a static pressure rise is obtained by allowing successive volumes of gas to be aspirated into and exhausted out of a closed space by means of the displacement of a moving member.

Displacement of a compressor—The volume displaced by the compressing element of the first stage per unit of time.

Disposable filter—A filter element intended to be discarded and replaced after one service cycle.

DOE—The U.S. Department of Energy.

Double acting compressor—A positive displacement compressor where the pistons compress gas on both the head-end and crank-end strokes.

Downstream—The portion of the flow stream which has already passed through the system or the portion of the system located after a filter or separator/filter.

Drag—Occurs when a valve does not close completely after popping and remains partly open until the pressure is further reduced.

Drain valve—A valve designed to remove surplus liquid from the compressed gas system (scrubbers, pulsation bottles, etc.).

Dripleg—A pipe extending downward from the bottom of the vessel to collect any condensation flow.

Dry bulb temperature—The ambient gas temperature as indicated by a standard thermometer.

Dry gas—Any gas or gas mixture that contains no water vapor and/or in which all of the constituents are substantially above their respective saturated vapor pressures at the existing temperature.

Dry unit (oil free)—Is one in which there is no liquid injection and/or liquid circulation for evaporative cooling or sealing.

Dynamic losses—Friction against duct walls, internal friction in the ga mass and direction variations will cause a speed reduction and are therefore called dynamic losses.

Dynamic type compressors—Machines in which air or gas is compressed by the mechanical action of rotating vanes or impellers imparting velocity and pressure to the flowing medium.

Dynamic viscosity (*Dynamic*)—The tangential force per unit area necessary to move one layer with respect to the other at unit velocity when maintained a unit distance apart by the fluid. Units are pound force sec per foot squared (lbf sec/ft$^{2)}$ or kilogram/meter sec (kg/m sec)

Durometer—This term refers to the hardness or softness of gaskets.

Dust cake—A layer of dust built up on an air filter.

Dust holding capacity—The amount of atmospheric dust which a filter will capture.

Effective area—The area (in sq inches) of the filter element that is exposed to the flow of air or fluid for effective filtering.

Efficiency—filter—Ability of a filter to remove particle matter from an air stream. Measured by comparing concentrate of material upstream and downstream of the filter. Typical particulate sizes range from .3 micron to 50 micron.

Efficiency- compression—Is the ratio of the theoretical work requirement to the actual work required to be performed on the gas for compression and delivery.

Efficiency- isothermal—The ratio of the theoretical work calculated on an isothermal basis to the actual work transferred to the gas during compression.

Efficiency mechanical—The ratio of the thermodynamic work requirement in the cylinder to actual brake horsepower requirement.

Efficiency—polytropic—The ratio of the polytropic compression energy transferred to the gas to the actual energy transferred to the gas.

Efficiency—volumetric—The ratio of actual capacity to piston displacement, stated as a percentage.

Element—The medium or material that does the actual filtering or separating. May be paper, wire mesh, special cellulose, inorganic plastic, or a combination.

Emulsibility—The ability of a non-water-soluble fluid to form an emulsion with water.

Emulsifier—Additive that promotes the formation of a stable mixture, or emulsion, of oil and water. Common emulsifiers are: metallic soaps, certain animal and vegetable oils, and various polar compounds.

Emulsion—Intimate mixture of oil and water, generally of a milky or cloudy appearance. Emulsions may be of two types: oil-in-water (where water is the continuous phase) and water-in-oil (where water is the discontinuous phase).

Energy audit—A survey that shows how much energy you use in your compressed air generation.

Energy conservation—Practices and measures that increase energy efficiency.

Energy kinetic—The energy a substance possesses by virtue of its motion or velocity.

Energy storage—The ability to convert energy into other forms, such as heat or chemical reaction, so that it can be retrieved for later use. Also the development, design, construction and operation of devices for storing energy until needed. Technology includes devices such as compressed gas.

Enthalpy—The sum of the internal and external energies.

Entropy—A measure of the unavailability of energy in a substance.

Environmental contaminant—All material and energy present in and around an operating system, such as dust, air moisture, chemicals, and thermal energy.

Evaporation—The escape of water molecules from a liquid to the gas phase at the surface of a body of water.

Exothermic—A term used to describe a chemical process in which heat is released.

Expanders—Turbines or engines in which gas expands, does work, and undergoes a drop in temperature.

Ferrography—An analytical method of assessing machine health by quantifying and examining ferrous wear particles suspended in the lubricant or hydraulic fluid.

Filter—A device that removes solid contaminants, such as dirt or metal particles, from a liquid or gas, or that separates one liquid from another, or a liquid from a gas.

Filter breather—A filtering unit for vented enclosures installed to prevent dirt and foreign matter from entering the enclosure. Also prevents oil loss by retaining oil droplets and draining the oil back to the sump.

Filter efficiency—Ability of a filter to remove particle matter from an air stream. Measured by comparing concentrate of material upstream and downstream of the filter. Typical particulate sizes range from .3 micron to 50 micron.

Filter element—The porous device which perform the actual process of filtration.

Filter housing—Something that covers or protects the filter assembly.

Filter separator—Filtering unit that separates solids and liquid droplets from gas (air). Widely used in removing oil from a gas or air.

Filtration—The physical or mechanical process of separating insoluble particulate matter from a fluid, such as air or liquid, by passing the fluid through a filter medium that will not allow the particulate to pass through it.

First law of thermodynamics—The amount of work done on or by a system is equal to the amount of energy transferred to or from the system.

Flange—A bolted rim used for attachment to another object.

Flash point—The lowest temperature to which oil must be heated under standardized test conditions to drive off sufficient inflammable vapor to flash when brought into contact with a flame. Flash points of petroleum based lubricants increase with increasing pressure.

Flexible mounting—Vibration isolation mount. Provides reductions in vibration transmission.

Flow—The volume of a substance passing a point per unit time (e.g., meters per second, gallons per hour, etc.).

Flow control valve—A valve that controls the flow of air that passes through the valve.

Flow diagram—A schematic flow sheet showing all controls involved with the system.

Flow meter—An instrument for measuring the amount of gas flow of a compressor. Measured in CFM.

Flow rate—The rate (in liters or gallons per minute, cubic meters or cubic feet per second, or other quantity per time unit)

Flushing—A circulation process designed to remove contamination.

Forced draft fan—A fan that generates (by pushing) a flow of ambient air over the exterior of the finned pipes to dissipate the sensible heat.

Friction—Surface resistance to relative motion, which slows down movement and causes heat.

Full load—Achieved when the air compressor is running at full RPM with a fully opened inlet and discharge, delivering the maximum volume at the rated pressure.

Gay-Lussac's Law—States that if the mass and pressure of a gas are held constant than gas volume is directly proportional to the gas's temperature.

Gag—A device attached to a safety relief valve that prevents it from opening at the set pressure.

Galling—A form of wear in which seizing or tearing of intermittent contacting surfaces occurs.

Gas—One of three basic phases of matter (solid, liquid, gas).

Gas bypass—All or a portion of gas that is redirected upstream or downstream of a component to compensate for process flow demands.

Gas drying—The drying of compressed gases other than air. Equipment size, choice of materials and other specifications may be decidedly different for drying gases other than air because of the specific properties of the gases. Properties include specific gravity, specific heat, viscosity, thermal conductivity, explosive characteristics, toxicity, corrosion and others.

Gas laws—The behavior of perfect gases, or mixtures thereof, follows a set of laws. Boyle' law, Charle's law, Amonton's law, Dalton's law, Amagat's law, Avogadro's law, Poisson's law.

Gate valve—A type of valve in which the closing element (the gate) is a disc that moves across the stream in a groove or slot for support against pressure. A gate valve has relatively large full ports and a straight line flow pattern.

Gage—An instrument for measuring, testing, or registering.

Gage pressure—Is pressure as determined by most instruments and gages.

Glycol dehydration—A method of drying natural gas.

Governor—A engine or motor controller that regulates the fuel flow or amperage to the engine or motor to control the speed and power.

GPM—Gallons per minute

Guide—The portion of a valve used to guide the disc.

Guide vane—A stationary element that may be adjustable and which directs the flow medium to the inlet of an impeller.

Head adiabatic—The force (or pressure) that a adiabatic (No heat transferred to or from the fluid) compression process raises the gas from suction level to discharge level measured in inches of feet of water column.

Header—A distribution pipe.

Head polytropic—The force (or pressure) that a polytropic (heat transfer to/from the fluid can exists and the fluid may not behave as an ideal gas)compression process raises the gas from suction level to discharge level measured in inches of feet of water column.

Head pressure—The force (or pressure) that the compression process raises the gas from suction level to discharge level measured in inches of feet of water column. This term is also referred to as **Head.**

Heat capacity—The heat energy transferred from system A to system B.

Heat exchanger—A device in which two fluids, separated by a heat conducting barrier, transfer heat energy from one fluid to the other.

Heat recovery—Recovering and utilizing the heat content of a fluid.

Horsepower (HP)—A unit of work equal to 33,000 foot pounds per minute, 550 foot pounds per second, or 746 Watts.

Horsepower brake (BHP)—The horsepower output from the driver and input to the compressor shaft. Also referred to as Shaft Horsepower.

Horsepower gas—The actual work required to compress and deliver a given gas quantity, including all thermodynamic, leakage and fluid friction losses.

Horsepower ideal—The horsepower required to isothermally compress gas.

Horsepower indicated—The horsepower calculated in positive displacement compressors using compressor-indicator diagrams.

Horsepower peak—The maximum power available from a driver or required by a compressor.

Horsepower theoretical—The horsepower required to adiabatically compress gas.

Hot gas bypass valve—A valve which connects the compressor discharge to the compressor suction and adjusted so as to maintain a specific pressure on the suction side by controlled bypass or to regulate compressor throughput.

Humidity—The moisture content of air.

Humidity specific—The weight of water vapor in the air vapor mixture per pound of dry air.

Humidity relative—The relative humidity of air vapor mixture is the ratio of the partial pressure of the vapor to the vapor saturation pressure at the dry bulb temperature of the mixture.

Hydrocarbons—Chemicals containing carbon and hydrogen.

ICFM—Gas flow in cubic feet per minute at the compressor inlet pressure & temperature conditions

I.D.—Pipe or vessel inside diameter.

Ideal gas—A gas that follows the perfect gas laws without deviation.

IGV—Inlet guide-vane valve. Valve(s) installed at the compressors inlet.

Immiscible—Fluids that are incapable of being mixed or blended together.

Impeller—The rotating element(s) of a dynamic compressor that imparts energy to the flowing medium by means of centrifugal force.

Inches of water—A pressure measurement – for example 407.258 in of water (at 60 ^{0}F) = 1 Atmosphere = 14.696 psia.

Indicated power—Power calculated from a compressor-indicator diagram.

Indicator card—A pressure-volume diagram for a compressor or engine cylinder produced directly from measurements made using of a device called the engine/compressor indicator.

Inducer—A curved inlet section on an impeller.

Inerting—Process of purging a compressor or a compartment with a non-reactive such **Inert gas**—A gas that has little or no heating value.

Inertia forces—Forces generated by the moving parts (such as pistons, rods, crossheads, connecting rods) with in the compressor and act on the compressor case and the support system.

Influent—The fluid entering a component.

Inlet pressure—The total pressure measured at the inlet flange of a compressor.

Inlet temperature—The temperature measured at the inlet flange of the compressor.

Inlet throttle—A compressor control device (valve) that controls the operation of the compressor.

Instrument air—Compressed air (usually dry and free from contaminants) used specifically with pneumatic instruments and controls.

Intercooler—Heat exchangers for removing the heat of compression between compressor stages.

Internal energy—Energy which a substance possesses because of the motion and configuration of its atoms, molecules, and subatomic particles.

International Organization for Standardization—ISO.

Irreversible process—A process in which the system and/or surroundings cannot be returned to their original state.

Isentropic compression—Compression at constant entropy

Isentropic efficiency—The ratio of isentropic work to actual work

ISO—International Organization for Standardization.

Isobar—Constant pressure.

Isochor—Constant volumn.

Isotherm—Constant temperature.

Isothermal compression—Compression where the temperature of a gas remains constant.

Isothermal efficiency—The ratio of the isothermal power consumption to shaft power input.

Joule—The amount of work done by a force of one newton moving an object through a distance of one meter. One joule is equal to one WATT—second or 0.737 foot pounds.

Joule effect—The conversion of mechanical, electrical or magnetic energy into heat.

Joule Thompson effect—The temperature change that occurs when a non-ideal gas suddenly expands from a high pressure to a low pressure. The ratio of the change in temperature divided by the change in pressure is known as the Joule-Thomson coefficient.

Journal—That part of a shaft or axle that rotates in a stationary fixed or tilt-pad bearing.

Journal bearing—A fixed two-piece or tilt-pad cylinder surrounding the shaft (journal) and supplied with a fluid lubricant. The fluid is the medium that supports the shaft preventing metal-to-metal contact.

Kelvin (K)—Pertains to the absolute scale of temperature in which the degree intervals are equal to those of the Celsius scale and in which the triple point of water has the value 273.16 Kelvin.

Kilowatt (kW)—A unit of power equal to 1,000 watts.

Kilowatt hour (kWh)—A unit of work done in one hour at the rate of 1,000 watts.

Kinematic viscosity—Is the ratio of the absolute or dynamic viscosity to density.

Kinetic energy—The energy of a body or a system with respect to the motion or velocity of the body or system or of the particles in the system.

Knock out—A term used to describe the condensate flow rate. It can be expressed as gallons per minute (GPM) or pounds mass per minute (lb_m/min).

kPa—Kilopascal; a metric measure of pressure based on force per unit area. (1 kPa = 4.01 inches of water).

kW—Kilowatt—A unit of power equal to 1,000 watts.

kWh—Kilowatt hour—A unit of work being done in one hour.

Lift—The distance between valve plate and the seating surfaces when the valve goes full open.

Liquid ring rotary compressor—A liquid piston compressor is a rotary compressor in which a vaned rotor revolves in an elliptical casing, with the rotor spaces sealed by a ring of liquid rotating with it inside the casing.

LNG—Abbreviation for Liquefied Natural Gas.

Load Electrical—Electric power consumed.

Load Mechanical—Mechanical or shaft power consumed.

Load factor—Ratio of the design compressor load to the maximum rated load of the compressor.

Lobe—A rotating element of a blower.

Lubrication—A material (lubricant) used between moving parts of machinery to reduce friction and carry away heat.

Mach number—The ratio of the actual velocity or speed to the speed of sound in the surrounding medium.

Manifold—A common chamber with multiple inlets and

Man way—An inspection cover or port in a vessel or tank.

MAWP—Maximum allowable working pressure.

Maximum operating pressure—The highest operating pressure the system or component is designed to withstand.

Mechanical efficiency—The ratio of work or power input to work or power output1.

Mesh size—Mesh is the number of openings in a square inch of screen or sieve.

Micron—Micrometer or one millionth of a meter; micron is sometimes represented in filtration by the Greek letter μ (mu). A micron is 0.000039". Contaminant particles are measured by micron size and count.

MMCFD—Millions of cubic feet per 24 hour.

Modulating control—Controlling a compressor at various speeds, throughput, pressure or temperature.

Moisture separator—A unit designed to separate condensate from the compressed air stream.

Moisture trap—A device designed to enable accumulated liquids to be held for draining from a vessel.

Mole—That mass of a substance which is numerically equal to its molecular weight

MSDS—Materials Safety Data Sheets

Multistage axial compressor—A machine having two or more impellers operating in series on a single shaft and in a single casing.

Multistage centrifugal compressor—A machine having two or more impellers operating in series on a single shaft and in a single casing.

Natural frequency—The natural frequency, or resonant frequency, is the frequency at which an element will vibrate when excited by a one time force.

N.B.—National Board of Boiler and Pressure Vessel Inspectors.

NEC—National Electrical Code

Negative pressure—A pressure below that of the existing atmospheric pressure taken as a zero reference.

NEMA—National Electrical Manufacturers Association.

NFPA—National Fluid Power Association.

No load—The air compressor continues to run, usually at full RPM, but no air is delivered because the inlet is either closed off or modified, not allowing inlet air to be trapped.

Nominal efficiency—An arbitrary filter efficiency rating.

Noncondensables—Are those constituents in the suction gas that cannot be condensed to a liquid ,, and removed from the gas stream, with the cooling medium available.

Noncooled compressor cylinders—Reciprocating type compressor cylinders, with low compression ratios, used mainly in oil and gas field applications.

Noncorrosive gas—Is one that does not attack normal materials of construction.

Nonlubricated compressor—A compressor designed to compress air or gas without contaminating the flow with lubricating oil.

Nozzle—A mechanical device designed to control the characteristics of a fluid as it exits (or enters) an enclosed chamber or pipe.

NPT—National Pipe Thread standard. A description of a specific pipe thread

NPTT—National Pipe Thread tapered. A description of a specific pipe thread

NTP—Normal Temperature and Pressure. This is based upon a temperature of 32 deg F (0 deg C) and a pressure of 1 bara.

O.D.—A measurement. Outside diameter.

OEM—Original equipment manufacturer.

Oil aerosol—A suspension of fine solid or liquid particles in a gas.

Oil free—The term generally applies to the condition of the air or gas either when it leaves the compressor, or after filtration.

Oil system—A arrangement of components such as pump, sump, separator, cooler and filter.

Operating pressure—The pressure at which a combination of components is maintained in normal operation.

Orifice—A restriction placed in a pipe line to provide a means of controlling or measuring flow.

OSHA—Occupational Safety and Health Administration.

Overhung type centrifugal compressor—A axial inlet compressor with the impeller or impellers mounted on an extended shaft.

Particulate type filter—A device designed to remove solids, such as dirt, scale, rust and other contaminants from the gas.

Pedestal type centrifugal compressor—A single or dual inlet compressor with the impeller or impellers mounted on a shaft between two bearings.

Performance curve—A plot of expected operating characteristics (e.g.., discharge pressure versus inlet capacity, shaft horsepower versus inlet capacity, efficiency).

pH—Measure of alkalinity or acidity in water and water containing fluids.

Pinion—The smaller of two mating or meshing gears; can be either the driving or the driven gear.

Piston displacement—Net volume actually displaced by the compressor piston when traveling from crank end to head end.

Pitot tube—A sensing device used to measure total pressures in a fluid stream.

Pleated filter—A filter element whose medium consists of a series of uniform folds.

Pneumatic power—The power of compressed air.

Pneumatic tools—Tools that operate by air pressure.

Positive displacement compressors—Compressors in which successive volumes of air or gas are confined within a closed space, and compressed.

Potential energy—The energy a substance possesses because of its relative elevation.

Pounds per square inch—PSI—Pounds per square inch.

Pour point—The temperature at which oil begins to flow under prescribed conditions.

PPB—A measurement. Parts Per Billion

PPBV—A measurement. Parts Per Billion by Volume.

PPM—A measurement. Parts per million

Precooler—A heat exchanger located immediately upstream of the compressor inlet.

Pressure—Force per unit area, usually expressed in pounds per square inch (PSI) or BAR.

Pressure absolute—The total pressure measured from absolute zero (i.e., from an absolute vacuum).

Pressure cracking—The pressure at which a pressure operated valve begins to open.

Pressure critical—The saturation pressure at the critical temperature.

Pressure dew point—The temperature at which moisture begins to condense in a compressed air system.

Pressure discharge—The total gas pressure (static plus velocity) at the discharge port of the compressor.

Pressure drop—Resistance to flow. Defined as the difference in pressure upstream and downstream.

Pressure rated—The qualified operating pressure which is recommended for a component or a system by the manufacturer.

Pressure regulating valve—A valve which enables pressure to be reduced, or kept constant at a desired level.

Pressure relief device—A device actuated by inlet static pressure and designed to open during an abnormal condition to prevent a rise of internal pressure in excess of a specified value.

Pressure total—The pressure that would be produced by stopping a moving stream of liquid or gas.

Preventive maintenance—Also known as PM, maintenance performed according to a fixed schedule involving the routine repair and replacement of machine parts and components.

Pseudo critical pressure—The saturation pressure at the critical temperature. It is the highest vapor pressure that the liquid can exert.

PSI—Pounds per square inch.

PSIA—Pounds per square inch, absolute.

PSID—Pounds per square inch, differential.

PSIG—Pounds per square inch, gauge. Pressure indicated by a pressure gauge.

Psychrometry—The relationship of the properties of air and water vapor mixtures in the atmosphere.

Pulsation bottle or damper—A receiver fitted on the inlet or discharge of a reciprocating compressor. The device is designed to remove resonances from the compressor piping thereby reducing noise and stresses.

Purging—Process of expelling an unwanted gas or liquid from a system through the introduction of a different gas until the last remnants of unwanted gas have been removed.

Receivers—Tanks used for the storage of air discharged from compressors.

Reciprocating compressors—Machines in which the compressing element is a piston having a reciprocating motion in a cylinder.

Reduced pressure—The ratio of the absolute pressure of a gas to its critical pressure.

Reduced temperature—The ratio of the absolute temperature of a gas to its critical temperature.

Regulator—An automatic or manual device designed to control pressure, flow or temperature.

Relative clearance volume—The ratio of clearance volume to the volume swept by the compressing element.

Relative humidity—The ratio of the actual water-vapor partial pressure to its saturation pressure at the same temperature. (considered only with atmospheric air).

Relative vapor pressure—The ratio of the vapor pressure to the saturated vapor pressure at the temperature considered.

Relief valve—A spring loaded pressure relief valve actuated by the static pressure upstream of the valve.

Reversible process—An ideal process, which once having taken place, can be reversed and in so doing leave no change in either system or surroundings

Reynold's number—The numerical value of a dimensionless combination of four variables (pipe diameter, density and viscosity of the flowing fluid and the velocity of flow) may be considered to be the ratio of the dynamic forces of mass flow to the shear stress due to viscosity.

Rings—Circular metallic elements that ride in the grooves of a compressor piston and provide sealing.

Rolling element—A type of anti-friction bearing in lightweight compressors.

Rotary blowers—A compressor belonging to the group of positive displacement rotary compressors.

Rotor—The rotating element of a compressor composed of the impeller(s), shaft, seals and may include shaft sleeves and a thrust balance piston.

SAE—Society of Automotive Engineers.

Safety valve—A device that limits fluid (liquid and gaseous) pressures beyond a preset level.

Safety relief valve—An automatic pressure relieving valve actuated by the static pressure of the system, which opens in proportion to the increase in pressure over the opening pressure.

Saturated vapor pressure—The pressure existing at the saturation temperature in a closed vessel containing a liquid and the vapor.

Saturation—Occurs when the vapor is at the dew point or saturation temperature corresponding to its partial pressure.

SCFM—Standard cubic feet per minute.

Screw compressor—A positive displacement rotary compressor.

Sea level—This is the designation for the average ISO pressure = 14.69 psia.

Seals—Devices used between rotating and stationary parts to eliminate or minimized leakage between areas of unequal pressures.

Second law of thermodynamics—A process can only proceed in a certain direction but not in the opposite direction (heat cannot pass from a colder body to a hotter body).

Set pressure—The gauge pressure at which a safety valve visibly and audibly opens or a setting at which a relief valve opens.

Shaft—The part of the rotating element on which the rotating parts are mounted and by means of which energy is transmitted from the prime mover.

Shaft input—The power required at the compressor drive shaft.

SI—Systeme International. The international system of unit measurement.

Silencer—A device fitted at a valves outlet to reduce or dampen the noise produced when the valve vents or opens.

Single acting—When a piston compresses gas only in one direction.

Single stage compressors—Compressors in which gas is compressed in each cylinder from initial intake pressure to final discharge pressure.

Single stage centrifugal compressors—Compressors having only one impeller.

Sleeve—A type of hydrodynamic journal bearing.

Slip—The internal leakage within a rotary compressor representing the partially compressed gas that is not delivered.

Sole plate—A metallic pad, usually embedded in concrete, on which the compressor feet are mounted.

Sonic flow—The point (speed of sound) at which air flow through an orifice can not increase regardless of pressure drop.

SPC—Specific Power Consumption.

Specific energy requirement—The shaft input per unit of compressor capacity.

Specific gravity—The ratio of the density of the material to the density of water at a specified temperature.

Specific heat—The quantity of heat required to raise the temperature of a unit weight of a substance by one degree.

Specific humidity—The weight of water vapor in an air-vapor mixture per pound of dry air.

Specific power—A measure of compressor efficiency, usually in the form of bhp/100 acfm or acfm/bhp

Specific volume—Is the volume of a given gas divided by its weight.

Specific weight—The weight of a gas per unit volume

Speed—The number of revolutions per minute of the compressor shaft.

Speed of sound—The rate at which sound waves travel.

Stages—Steps in the compression of a gas, In reciprocating compressors, each stage usually requires a separate cylinder, in dynamic compressors, each requires a separate rotor disc.

Stem—The rod connecting the disc to the lever on a valve.

Suction pressure—This is the pressure found on the suction side of a compressor.

Surge—The reversal of flow within a dynamic compressor resulting from insufficient head and flow

Surge limit—In a dynamic compressor, surge limit is the capacity below which the compressor operation becomes unstable.

Synthetic lubricant—A lubricating oil made from synthetic based materials.

Temperature—The physical property of a body that lies behind the common notion of hot and cold.

Temperature absolute—The temperature of a body relative to the absolute zero temperature, at which point the volume of an ideal gas theoretically becomes zero. (Fahrenheit scale is minus 459.67°F/Celsius scale is minus 273.15°C).

Temperature discharge—The temperature of the medium at the discharge flange of the compressor.

Temperature inlet—The temperature of the medium at the inlet flange of the compressor.

Temperature static—The temperature at a stagnation (speed of the fluid is zero) point in the fluid flow. Also known as the stagnation temperature.

Thermodynamics first law of—The amount of work done on or by a system is equal to the amount of energy transferred to or from the system.

Thermodynamics second law of—Heat cannot, of itself, pass from a colder to a hotter body.

Thrust balancing piston—The part of a rotating element that counteracts the thrust developed by the impellers.

Tilting pad—A type of hydrodynamic journal bearing in a compressor.

Total package input power—The total electrical power input to a electric motor driven compressor, including drive motor, cooling fan, auxiliary motors, controls, etc.

Torque—Torsional moment or couple. It usually refers to the driving couple of a machine or motor.

Two stage compressor—Compressors in which gas is compressed from an initial pressure to an intermediate pressure in one stage and than compressed from the intermediate stage to the final stage.

Two step control—Load/unload control system that tries to maximizes compressor efficiency by matching delivery to demand. Basically the compressor is operated at full load or idle.

UL—Underwriters Lab.

Ultrasonic leak detector—An instrument designed to detect the ultrasonic emissions and convert them to audible signal.

UNC—Thread—Unified national coarse.

UNF—Thread—Unified national fine.

Unload—A condition where the compressor runs at full RPM but no gas is delivered because the inlet is closed off.

Unloaded horsepower—The power that is consumed to overcome the frictional losses when operating in an unloaded condition.

Vacuum pumps—Compressors that operate with an intake pressure below atmospheric and discharge pressure usually atmospheric or slightly higher.

Valves—Devices with passages for directing flow into alternate paths or restricting flow along a given path.

Vane compressor—A single shaft, positive displacement rotary compressor.

Vapor—Fine separated particles floating in the air and clouding it. A substance in the gaseous state.

Vaporization—The process of changing from a liquid to a gaseous state

Vapor pressure—The pressure exerted by a vapor confined within a given space.

V belt drive—A drive arrangement for power transmission to compressors.

Venturi—A reduction in fluid pressure that results when a fluid flows through a constricted section of pipe.

Viscosity—A measure of the resistance of a fluid which is being deformed by either shear stress or external stress.

Viscosity index (VI)—A measure of the rate of change of viscosity with temperature.

Volumetric efficiency—The ratio and percent of the actual delivered capacity (measured at inlet temperature, pressure and gas composition) to the piston displacement.

Volute—A stationary, spirally shaped passage that converts velocity head to pressure.

Water cooled compressors—Compressors cooled by water/glycol circulated through jackets surrounding the cylinders or casings.

Water solubility—The ability or tendency of a substance to blend uniformly with water.

Wet bulb temperature—The temperature recorded by a thermometer whose bulb has been covered with a wetted wick and whirled on a sling psychrometer. Taken with the dry bulb, it permits determination of relative humidity of the atmosphere.

Wet gas—A gas or gas mixture in which one or more of the constituents is at its saturated vapor pressure. The constituent at saturation pressure may or may not be water vapor.

Wicking—The vertical absorption of a liquid into a porous material by capillary forces.

WMW—Welded Minimum Wall tubing

WOG—WOG—A pressure rating in psi for valves. W. Water, O. Oil, G. Gas

Work—Energy in transition and is defined in units of Force times Distance. Work cannot be done unless there is movement.

Yoke—The portion of a safety/relief valve that surrounds the spring. The spring housing.

Appendix B3

Conversion Factors

TO OBTAIN	MULTIPLY	BY
bars	atmospheres	1.0133
cm Hg @ 32°F (0°C)	atmospheres	76.00
ft of water @ 40°F (4°C)	atmospheres	33.90
in. Hg @ 32°F (0°C)	atmospheres	29.92
in. Hg @ 60°F (15°C)	atmospheres	30.00
lb/sq ft.	atmospheres	2,116
lb/sq. in.	atmospheres	14.696
cm Hg @ 32°F (0°C)	bars	75.01
in. Hg @ 32°F (0°C)	bars	29.53
lb/sq ft.	bars	2,088
lb/sq in.	bars	14.50
ft-lb	Btu	778.26
Horsepower-hr	Btu	3.930×10^{-4}
joules	Btu	1055
kilogram calories	Btu	2.520×10^{-1}
kilowatt-hr	Btu	2.930×10^{-4}
grams	carats	2.0×10^{-1}
ounces	carats	7.055×10^{-3}
ft.	centimeters	3.281×10^{-2}
in.	centimeters	3.937×10^{-1}
kilogram/hr m	centipoise	3.60
lb/sec ft	centipoise	6.72×10^{-4}
Poise	centipoise	1×10^{-2}
ft of water @ 40°F (4°C)	cm Hg	4.460×10^{-1}
in. H$_2$O @ 40°F (4°C)	cm Hg	5.354
kilogram/sq m.	cm Hg	135.95
lb/sq. in.	cm Hg	1.934×10^{-1}
lb/sq./ft.	cm Hg	27.85
ft/sec	cm/sec	3.281×10^{-2}
mph	cm/sec	2.237×10^{-2}

cu in.	cu cm (dry)	6.102×10^{-2}
gal U.S.	cu cm (liquid)	2.642×10^{-4}
liters	cu cm (liquid)	1×10^{-3}
cu cm	cu ft (dry)	2.832×10^{4}
cu in.	cu ft (dry)	1,728
liters	cu ft (liquid)	28.32
lb	cu ft H_2O	62.428
gal U.S.	cu ft liquid)	7.481
cu m/min	cu ft/min	2.832×10^{-2}
gallons/hr	cu ft/min	4.488×10^{2}
liters/sec	cu ft/min	4.719×10^{-1}
cu cm	cu in.	16.39
gal U.S.	cu in.	4.329×10^{-3}
liters	cu in.	1.639×10^{-2}
quarts	cu in.	1.732×10^{-2}
cu ft.	cu meters	35.31
cu in.	cu meters	61,023
gas U.S.	cu meters	264.17
liters	cu meters	1.0×10^{3}
meters	ft	3.048×10^{-1}
cm/sec	ft/min	5.080×10^{-1}
km/hr	ft/min	1.829×10^{-2}
mph	ft/min	1.136×10^{-2}
cm/sec	ft/sec	30.48
gravity	ft/sec	32.1816
km/hr	ft/sec	1.097
knots	ft/sec	5.925×10^{-1}
mph	ft/sec	6.818×10^{-1}
Btu	gram-calories	3.969×10^{-2}
joules	gram-calories	4.184
kg/m	gram/cm	1×10^{-1}
lb/ft	gram/cm	6.721×10^{-2}
lb/in.	gram/cm	5.601×10^{-3}
kg/cu m	grams/cu cm	1000
lb/cu ft	grams/cu cm	62.43
slugs	grams/cu cm	1.940
Btu/hr	horsepower	2,546
Btu/sec	horsepower	7.073×10^{-1}
ft-lb/min	horsepower	33,000

ft-lb/sec	horsepower	550
kilowatts	horsepower	7.457×10^{-1}
m-kg/sec	horsepower	76.04
metric hp	horsepower	1.014
Btu-sec	horsepower, metric	6.975×10^{-1}
horsepower	horsepower, metric	9.863×10^{-1}
kilowatts	horsepower, metric	7.355×10^{-1}
m-kg/sec	horsepower, metric	75.0
Btu	horsepower-hours	2.545×10^3
ft-lb	horsepower-hours	1.98×10^6
kilowatt-hours	horsepower-hours	7.457×10^{-1}
m-kg	horsepower-hours	2.737×10^5
cm Hg @ 32°F (0°C)	in H_2O @ 40°F (4°C)	1.868×10^{-1}
in Hg @ 32°F (0°C)	in H_2O @ 40°F (4°C)	7.355×10^{-2}
kg/sq m	in H_2O @ 40°F (4°C)	25.40
lb/sq ft.	in H_2O @ 40°F (4°C)	5.202
lb/sq in.	in H_2O @ 40°F (4°C)	3.613×10^{-2}
atmospheres	in H_2O @ 60°F (15°C)	2.455×10^{-3}
in Hg @ 32°F (0°C)	in H_2O @ 60°F (15°C)	7.349×10^{-2}
atmospheres	in Hg @ 32°F (0°C)	3.342×10^{-2}
ft H_2O 40°F (4°C)	in Hg @ 32°F (0°C)	1.133
in. H_2O @ 40°F (4°C)	in Hg @ 32°F (0°C)	13.60
kg/sq m	in Hg @ 32°F (0°C)	3.453×10^2
lb/sq ft.	in Hg @ 32°F (0°C)	70.73
lb/sq in.	in Hg @ 32°F (0°C)	4.912×10^{-1}
cm	in.	2.54
Btu	Joules	9.483×10^{-4}
ft-lb	Joules	7.376×10^{-1}
hp-hr	Joules	3.725×10^{-7}
kg calories	Joules	2.389×10^{-4}
kg m	Joules	1.020×10^{-1}
watt-hr	Joules	2.778×10^{-4}
grams/cu cm	kg/cu m	1×10^{-3}
lb/cu ft	kg/cu m	62.43×10^{-3}
in H_2O @ 40°F (4°C)	kg/sq cm	3.28×10^{-7}
in Hg @ 32°F (0°C)	kg/sq cm	28.96
lb/sq ft.	kg/sq cm	2.048×10^3
lb/sq in.	kg/sq cm	14.22
Btu	kilogram-calories	3.9680

ft-lb	kilogram-calories	3087
m-kg	kilogram-calories	4.269×10^2
grams	kilograms	1×10^3
lb	kilograms	2.205
oz	kilograms	35.27
cm	kilometers	1×10^5
ft	kilometers	3.281×10^3
miles	kilometers	6.214×10^{-1}
nautical miles	kilometers	5.400×10^{-1}
Btu/sec	kilowatts	9.485×10^{-1}
ft-lb/sec	kilowatts	7.376×10^2
horsepower	kilowatts	1.341
kg calories/sec	kilowatts	2.389×10^{-1}
ft/sec, fps	km/hr	9.113×10^{-1}
knots	km/hr	5.396×10^{-1}
m/sec	km/hr	2.778×10^{-1}
mph	km/hr	6.214×10^{-1}
ft/sec, fps	knots	1.688
km/hr	knots	1.853
m/sec	knots	5.148×10^{-1}
mph	knots	1.151
grains	lb, avdp	7000
grams	lb, avdp	453.6
ounces	lb, avdp	16.0
poundals	lb, avdp	32.174
slugs	lb, avdp	3.108×10^{-2}
kg/cu m	lb/cu ft	16.02
grams/cu cm	lb/cu in.	27.68
lb/cu ft	lb/cu in.	1728
atmospheres	lb/sq in.	6.805×10^{-2}
ft H_2O @ 4°C	lb/sq in.	2.307
in Hg @ 0°C	lb/sq in.	2.036
kg/sq m	lb/sq in.	7.031×10^2
cu cm	liters	1×10^3
cu ft	liters	3.532×10^{-2}
cu in	liters	61.03
gal Imperial	liters	2.200×10^{-1}
gal U.S.	liters	2.642×10^{-1}
quarts	liters	1.057

ft-lb	meter-kilogram	7.233
joules	meter-kilogram	9.807
ft/sec, fps	meter/sec	3.281
km/hr	meter/sec	3.600
miles/hr	meter/sec	2.237
ft	meters	3.281
in.	meters	39.37
miles	meters	6.214×10^{-4}
in.	microns	3.937×10^{-5}
ft	miles	5280
kilometers	miles	1.609
nautical mile	miles	8.690×10^{-1}
in Hg @ 0°C	milibars	2.953×10^{-2}
ft/sec, fps	mph	1.467
km/hr	mph	1.609
knots	mph	8.690×10^{-1}
m/sec	mph	4.470×10^{-1}
ft	nautical mile	6076.1
m	nautical mile	1852
miles	nautical mile	1.151
grains	ounces, avdp	4.375×10^{2}
grams	ounces, avdp	28.35
lb, avdp	ounces, avdp	6.250×10^{-2}
cu cm	ounces. fluid	29.57
cu in.	ounces. fluid	1.805
degree (arc)	radians	57.30
degrees/sec	radians/sec	57.30
rev/min, rpm	radians/sec	9.549
rev/sec	radians/sec	15.92×10^{-2}
radians/sec	rev/min. rpm	1.047×10^{-1}
radians	revolutions	6.283
lb	slug	32.174
sq ft	sq cm	1.076×10^{-3}
sq in.	sq cm	1.550×10^{-1}
sq cm	sq ft	929.0
sq in.	sq ft	144.0
sq cm	sq in.	6.452
Btu/sec	watts	9.485×10^{-4}

Appendix C

Gas Properties and Behavior

This information is reprinted from the Gas Processors Suppliers Association Engineering Data Book Courtesy of The Gas Processors Suppliers Association.

The Physical Constants of Hydrocarbons, Compressibility Charts and Pressure-Enthalpy Diagrams have been reprinted in both English and SI Units.

Physical Constants of Hydrocarbons—English units
Compressibility Factors of Gases—English Units

Physical Constants of Hydrocarbons—SI units
Compressibility Factors of Gases—SI Units

Pressure-Enthalpy Diagrams

English Units
    Enthalpy—Pure Components
    Carbon Dioxide
    Methane
    Propane
    Ethane
    Ethylene
    Propylene
    Nitrogen
    Iso-Pentane

SI Units
    Carbon Dioxide
    Methane
    Propane
    Ethane
    Ethylene
    Propylene
    Nitrogen
    Iso-Pentane
    n-Butane
    n-Pentane
    Iso-Butane
    Water

Physical Constants of Hydrocarbons[1]

No.	Compound	Formula	Molecular weight	Boiling point °F., 14.696 psia	Vapor pressure, 100°F., psia	Freezing point, °F., 14.696 psia	Critical constants		
							Pressure, psia	Temperature, °F.	Volume, cu ft/lb
1	Methane	CH_4	16.043	−258.69	(5000)	−296.46^d	667.8	−116.63	0.0991
2	Ethane	C_2H_6	30.070	−127.48	(800)	−297.89^d	707.8	90.09	0.0788
3	Propane	C_3H_8	44.097	−43.67	190.	−305.84^d	616.3	206.01	0.0737
4	n−Butane	C_4H_{10}	58.124	31.10	51.6	−217.05	550.7	305.65	0.0702
5	Isobutane	C_4H_{10}	58.124	10.90	72.2	−255.29	529.1	274.98	0.0724
6	n−Pentane	C_5H_{12}	72.151	96.92	15.570	−201.51	488.6	385.7	0.0675
7	Isopentane	C_5H_{12}	72.151	82.12	20.44	−255.83	490.4	369.10	0.0679
8	Neopentane	C_5H_{12}	72.151	49.10	35.9	2.17	464.0	321.13	0.0674
9	n−Hexane	C_6H_{14}	86.178	155.72	4.956	−139.58	436.9	453.7	0.0688
10	2−Methylpentane	C_6H_{14}	86.178	140.47	6.767	−244.63	436.6	435.83	0.0681
11	3−Methylpentane	C_6H_{14}	86.178	145.89	6.098		453.1	448.3	0.0681
12	Neohexane	C_6H_{14}	86.178	121.52	9.856	−147.72	446.8	420.13	0.0667
13	2,3−Dimethylbutane	C_6H_{14}	86.178	136.36	7.404	−199.38	453.5	440.29	0.0665
14	n−Heptane	C_7H_{16}	100.205	209.17	1.620	−131.05	396.8	512.8	0.0691
15	2−Methylhexane	C_7H_{16}	100.205	194.09	2.271	−180.89	396.5	495.00	0.0673
16	3−Methylhexane	C_7H_{16}	100.205	197.32	2.130		408.1	503.78	0.0646
17	3−Ethylpentane	C_7H_{16}	100.205	200.25	2.012	−181.48	419.3	513.48	0.0665
18	2,2−Dimethylpentane	C_7H_{16}	100.205	174.54	3.492	−190.86	402.2	477.23	0.0665
19	2,4−Dimethylpentane	C_7H_{16}	100.205	176.89	3.292	−182.63	396.9	475.95	0.0668
20	3,3−Dimethylpentane	C_7H_{16}	100.205	186.91	2.773	−210.01	427.2	505.85	0.0662
21	Triptane	C_7H_{16}	100.205	177.58	3.374	−12.82	428.4	496.44	0.0636
22	n−Octane	C_8H_{18}	114.232	258.22	0.537	−70.18	360.6	564.22	0.0690
23	Diisobutyl	C_8H_{18}	114.232	228.39	1.101	−132.07	360.6	530.44	0.0676
24	Isooctane	C_8H_{18}	114.232	210.63	1.708	−161.27	372.4	519.46	0.0656
25	n−Nonane	C_9H_{20}	128.259	303.47	0.179	−64.28	332.	610.68	0.0684
26	n−Decane	$C_{10}H_{22}$	142.286	345.48	0.0597	−21.36	304.	652.1	0.0679
27	Cyclopentane	C_5H_{10}	70.135	120.65	9.914	−136.91	653.8	461.5	0.059
28	Methylcyclopentane	C_6H_{12}	84.162	161.25	4.503	−224.44	548.9	499.35	0.0607
29	Cyclohexane	C_6H_{12}	84.162	177.29	3.264	43.77	591.	536.7	0.0586
30	Methylcyclohexane	C_7H_{14}	98.189	213.68	1.609	−195.87	503.5	570.27	0.0600
31	Ethylene	C_2H_4	28.054	−154.62	—	−272.45^d	729.8	48.58	0.0737
32	Propene	C_3H_6	42.081	−53.90	226.4	−301.45^d	669.	196.9	0.0689
33	1−Butene	C_4H_8	56.108	20.75	63.05	−301.63^d	583.	295.6	0.0685
34	Cis−2−Butene	C_4H_8	56.108	38.69	45.54	−218.06	610.	324.37	0.0668
35	Trans−2−Butene	C_4H_8	56.108	33.58	49.80	−157.96	595.	311.86	0.0680
36	Isobutene	C_4H_8	56.108	19.59	63.40	−220.61	580.	292.55	0.0682
37	1−Pentene	C_5H_{10}	70.135	85.93	19.115	−265.39	590.	376.93	0.0697
38	1,2−Butadiene	C_4H_6	54.092	51.53	(20.)	−213.16	(653.)	(339.)	(0.0649)
39	1,3−Butadiene	C_4H_6	54.092	24.06	(60.)	−164.02	628.	306.	0.0654
40	Isoprene	C_5H_8	68.119	93.30	16.672	−230.74	(558.4)	(412.)	(0.0650)
41	Acetylene	C_2H_2	26.038	−119^e	—	−114.^{d}	890.4	95.31	0.0695
42	Benzene	C_6H_6	78.114	176.17	3.224	41.96	710.4	552.22	0.0531
43	Toluene	C_7H_8	92.141	231.13	1.032	−138.94	595.9	605.55	0.0549
44	Ethylbenzene	C_8H_{10}	106.168	277.16	0.371	−138.91	523.5	651.24	0.0564
45	o−Xylene	C_8H_{10}	106.168	291.97	0.264	−13.30	541.4	675.0	0.0557
46	m−Xylene	C_8H_{10}	106.168	282.41	0.326	−54.12	513.6	651.02	0.0567
47	p−Xylene	C_8H_{10}	106.168	281.05	0.342	55.86	509.2	649.6	0.0572
48	Styrene	C_8H_8	104.152	293.29	(0.24)	−23.10	580.	706.0	0.0541
49	Isopropylbenzene	C_9H_{12}	120.195	306.34	0.188	−140.82	465.4	676.4	0.0570
50	Methyl Alcohol	CH_4O	32.042	148.1(2)	4.63(22)	−143.82(22)	1174.2(21)	462.97(21)	0.0589(21)
51	Ethyl Alcohol	C_2H_6O	46.069	172.92(22)	2.3(7)	−173.4(22)	925.3(21)	469.58(21)	0.0580(21)
52	Carbon Monoxide	CO	28.010	−313.6(2)	—	−340.6(2)	507.(17)	−220.(17)	0.0532(17)
53	Carbon Dioxide	CO_2	44.010	−109.3(2)	—	—	1071.(17)	87.9(23)	0.0342(23)
54	Hydrogen Sulfide	H_2S	34.076	−76.6(24)	394.0(6)	−117.2(7)	1306.(17)	212.7(17)	0.0459(24)
55	Sulfur Dioxide	SO_2	64.059	14.0(7)	88.(7)	−103.9(7)	1145.(24)	315.5(17)	0.0306(24)
56	Ammonia	NH_3	17.031	−28.2(24)	212.(7)	−107.9(2)	1636.(17)	270.3(24)	0.0681(17)
57	Air	N_2O_2	28.964	−317.6(2)	—	—	547.(2)	−221.3(2)	0.0517(3)
58	Hydrogen	H_2	2.016	−423.0(24)	—	−434.8(24)	188.1(17)	−399.8(17)	0.5167(24)
59	Oxygen	O_2	31.999	−297.4(2)	—	−361.8(24)	736.9(24)	−181.1(17)	0.0382(24)
60	Nitrogen	N_2	28.013	−320.4(2)	—	−346.0(24)	493.0(24)	−232.4(24)	0.0514(17)
61	Chlorine	Cl_2	70.906	−29.3(24)	158.(7)	−149.8(24)	1118.4(24)	291.(17)	0.0281(17)
62	Water	H_2O	18.015	212.0	0.9492(12)	32.0	3208.(17)	705.6(17)	0.0500(17)
63	Helium	He	4.003	—	—	—	—	—	—
64	Hydrogen Chloride	HCl	36.461	−121(16)	925.(7)	−173.6(16)	1198.(17)	124.5(17)	0.0208(17)

Physical Constants of Hydrocarbons[1]

Specific gravity 60°F./60°F. o,b	lb/gal * a (Wt in vacuum)	lb/gal * a,c (Wt in air)	Gal/lb Mole*	Temperature Coefficient of density*a	Pitzer acentric factor (18)	Compressibility factor of real gas, Z 14.696 psia, 60°F.	Specific gravity Air = 1*	cu ft gas/lb*	cu ft gas/gal liquid*	Cp, Btu/lb/°F. Ideal gas	Cp, Btu/lb/°F. Liquid	No.
0.3^i	2.5^i	2.5^i	6.4^i	—	0.0104	0.9981	0.5539	23.65	59.^i	0.5266	—	1
0.3564^h	2.971^h	2.962^h	10.12^h	—	0.0986	0.9916	1.0382	12.62	37.5^h	0.4097	0.9256	2
0.5077^h	4.233^h	4.223^h	10.42^h	0.00152^h	0.1524	0.9820	1.5225	8.606	36.43^h	0.3881	0.5920	3
0.5844^h	4.872^h	4.865^h	11.93^h	0.00117^h	0.2010	0.9667	2.0068	6.529	31.81^h	0.3867	0.5636	4
0.5631^h	4.695^h	4.686^h	12.38^h	0.00119^h	0.1848	0.9696	2.0068	6.529	30.65^h	0.3872	0.5695	5
0.6310	5.261	5.251	13.71	0.00087	0.2539	0.9549	2.4911	5.260	27.67	0.3883	0.5441	6
0.6247	5.208	5.199	13.85	0.00090	0.2223	0.9544	2.4911	5.260	27.39	0.3827	0.5353	7
0.5967^h	4.975^h	4.965^h	14.50^h	0.00104^h	0.1969	0.9510	2.4911	5.260	26.17^h	(0.3866)	0.554	8
0.6640	5.536	5.526	15.57	0.00075	0.3007	—	2.9753	4.404	24.38	0.3864	0.5332	9
0.6579	5.485	5.475	15.71	0.00078	0.2825	—	2.9753	4.404	24.15	0.3872	0.5264	10
0.6689	5.577	5.568	15.45	0.00075	0.2741	—	2.9753	4.404	24.56	0.3815	0.507	11
0.6540	5.453	5.443	15.81	0.00078	0.2369	—	2.9753	4.404	24.01	0.3809	0.5165	12
0.6664	5.556	5.546	15.51	0.00075	0.2495	—	2.9753	4.404	24.47	0.378	0.5127	13
0.6882	5.738	5.728	17.46	0.00069	0.3498	—	3.4596	3.787	21.73	0.3875	0.5283	14
0.6830	5.694	5.685	17.60	0.00068	0.3336	—	3.4596	3.787	21.57	(0.390)	0.5223	15
0.6917	5.767	5.757	17.38	0.00069	0.3257	—	3.4596	3.787	21.84	(0.390)	0.511	16
0.7028	5.859	5.850	17.10	0.00070	0.3095	—	3.4596	3.787	22.19	(0.390)	0.5145	17
0.6782	5.654	5.645	17.72	0.00072	0.2998	—	3.4596	3.787	21.41	(0.395)	0.5171	18
0.6773	5.647	5.637	17.75	0.00072	0.3048	—	3.4596	3.787	21.39	0.3906	0.5247	19
0.6976	5.816	5.807	17.23	0.00065	0.2840	—	3.4596	3.787	22.03	(0.395)	0.502	20
0.6946	5.791	5.782	17.30	0.00069	0.2568	—	3.4596	3.787	21.93	0.3812	0.4995	21
0.7068	5.893	5.883	19.39	0.00062	0.4018	—	3.9439	3.322	19.58	(0.3876)	0.5239	22
0.6979	5.819	5.810	19.63	0.00065	0.3596	—	3.9439	3.322	19.33	(0.373)	0.5114	23
0.6962	5.804	5.795	19.68	0.00065	0.3041	—	3.9439	3.322	19.28	0.3758	0.4892	24
0.7217	6.017	6.008	21.32	0.00063	0.4455	—	4.4282	2.959	17.80	0.3840	0.5228	25
0.7342	6.121	6.112	23.24	0.00055	0.4885	—	4.9125	2.667	16.33	0.3835	0.5208	26
0.7504	6.256	6.247	11.21	0.00070	0.1955	0.9657	2.4215	5.411	33.85	0.2712	0.4216	27
0.7536	6.283	6.274	13.40	0.00071	0.2306	—	2.9057	4.509	28.33	0.3010	0.4407	28
0.7834	6.531	6.522	12.89	0.00068	0.2133	—	2.9057	4.509	29.45	0.2900	0.4332	29
0.7740	6.453	6.444	15.22	0.00063	0.2567	—	3.3900	3.865	24.94	0.3170	0.4397	30
—	—	—	—	—	0.0868	0.9938	0.9686	13.53	—	0.3622	—	31
0.5220^h	4.352^h	4.343^h	9.67^h	0.00189^h	0.1405	0.9844	1.4529	9.018	39.25^h	0.3541	0.585	32
0.6013^h	5.013^h	5.004^h	11.19^h	0.00116^h	0.1906	0.9704	1.9372	6.764	33.91^h	0.3548	0.535	33
0.6271^h	5.228^h	5.219^h	10.73^h	0.00098^h	0.1953	0.9661	1.9372	6.764	35.36^h	0.3269	0.5271	34
0.6100^h	5.086^h	5.076^h	11.03^h	0.00107^h	0.2220	0.9662	1.9372	6.764	34.40^h	0.3654	0.5351	35
0.6004^h	5.006^h	4.996^h	11.21^h	0.00120^h	0.1951	0.9689	1.9372	6.764	33.86^h	0.3701	0.549	36
0.6457	5.383	5.374	13.03	0.00089	0.2925	0.9550	2.4215	5.411	29.13	0.3635	0.5196	37
0.658^h	5.486^h	5.470^h	9.86^h	0.00098^h	0.2485	(0.969)	1.8676	7.016	38.49^h	0.3458	0.5408	38
0.6272^h	5.229^h	5.220^h	10.34^h	0.00113^h	0.1955	(0.965)	1.8676	7.016	36.69^h	0.3412	0.5079	39
0.6861	5.720	5.711	11.91	0.00086	0.2323	(0.962)	2.3519	5.571	31.87	0.357	0.5192	40
0.615^k	—	—	—	—	0.1803	0.9925	0.8990	14.57	—	0.3966	—	41
0.8844	7.373	7.365	10.59	0.00066	0.2125	0.929(15)	2.6969	4.858	35.82	0.2429	0.4098	42
0.8718	7.268	7.260	12.68	0.00060	0.2596	0.903(21)	3.1812	4.119	29.94	0.2598	0.4012	43
0.8718	7.268	7.259	14.61	0.00054	0.3169	—	3.6655	3.574	25.98	0.2795	0.4114	44
0.8848	7.377	7.367	14.39	0.00055	0.3023	—	3.6655	3.574	26.37	0.2914	0.4418	45
0.8687	7.243	7.234	14.66	0.00054	0.3278	—	3.6655	3.574	25.89	0.2782	0.4045	46
0.8657	7.218	7.209	14.71	0.00054	0.3138	—	3.6655	3.574	25.80	0.2769	0.4083	47
0.9110	7.595	7.586	13.71	0.00057	—	—	3.5959	3.644	27.67	0.2711	0.4122	48
0.8663	7.223	7.214	16.64	0.00054	0.2862	—	4.1498	3.157	22.80	0.2917	(0.414)	49
0.796(3)	6.64	6.63	4.83	—	—	—	1.1063	11.84	78.6	0.3231^v(24)	0.594(7)	50
0.794(3)	6.62	6.61	6.96	—	—	—	1.5906	8.237	54.5	0.3323^v(24)	0.562(7)	51
0.801^m(8)	6.68^m	6.67^m	4.19^m	—	0.041	0.9995(15)	0.9671	13.55	—	0.2484(13)	—	52
0.827^h(6)	6.89^h	6.88^h	6.38^h	—	0.225	0.9943(15)	1.5195	8.623	59.5^h	0.1991(13)	—	53
0.79^h(6)	6.59^h	6.58^h	5.17^h	—	0.100	0.9903(15)	1.1765	11.14	73.3^h	0.238(4)	—	54
1.397^h(14)	11.65^h	11.64^h	5.50^h	—	0.246	—	2.2117	5.924	69.0^h	0.145(7)	0.325^h(7)	55
0.6173(11)	5.15	5.14	3.31	—	0.255	—	0.5880	22.28	114.7	0.5002(10)	1.114^h(7)	56
0.856^m(8)	7.14^m	7.13^m	4.06^m	—	—	0.9996(15)	1.0000	13.10	—	0.2400(9)	—	57
0.07^m(3)	—	—	—	—	0.000	1.0006(15)	0.0696	188.2	—	3.408(13)	—	58
1.140(25)	9.50^m	9.49^m	3.37^m	—	0.0213	—	1.1048	11.86	—	0.2188(13)	—	59
0.810(26)	6.75^m	6.74^m	4.15^m	—	0.040	0.9997(15)	0.9672	13.55	—	0.2482(13)	—	60
1.414(14)	11.79	11.78	6.01	—	—	—	2.4481	5.352	63.1	0.119(7)	—	61
1.000	8.337	8.328	2.16	—	0.348	—	0.6220	21.06	175.6	0.4446(13)	1.0009(7)	62
—	—	—	—	—	—	—	—	—	—	—	—	63
0.8558(14)	7.135	7.126	5.11	0.00335*	—	—	1.2588	10.41	74.3	0.190(7)	—	64

Physical Constants of Hydrocarbons[1]

No.	Compound	Calorific value, 60°F.* Net — Btu/cu ft Ideal gas, 14.696 psia (20)*	Gross — Btu/cu ft Ideal gas, 14.696 psia (20)*	Btu/lb liquid (wt in vacuum)	Btu/gal liquid*	Heat of vaporization, 14.696 psia at boiling point, Btu/lb	Refractive index, nD 68°F.	Air required for combustion ideal gas* cu ft/cu ft	Flammability limits, vol % in air mixture — Lower	Higher	ASTM octane number — Motor method D-357	Research method D-908
1	Methane	909.1	1009.7	—	—	219.22	—	9.54	5.0	15.0	—	—
2	Ethane	1617.8	1768.8	—	—	210.41	—	16.70	2.9	13.0	+.05[f]	+1.6[j,f]
3	Propane	2316.1	2517.5	21513	91065	183.05	—	23.86	2.1	9.5	97.1	+1.8[j,f]
4	n–Butane	3010.4	3262.1	21139	102989	165.65	1.3326[h]	31.02	1.8	8.4	89.6[j]	93.8[j]
5	Isobutane	3001.1	3252.7	21091	99022	157.53	—	31.02	1.8	8.4	97.6	+.10[j,f]
6	n–Pentane	3707.5	4009.6	20928	110102	153.59	1.35748	38.18	1.4	8.3	62.6[j]	61.7[j]
7	Isopentane	3698.3	4000.3	20889	108790	147.13	1.35373	38.18	1.4	(8.3)	90.3	92.3
8	Neopentane	3682.6	3984.6	20824	103599	135.58	1.342[h]	38.18	1.4	(8.3)	80.2	85.5
9	n–Hexane	4403.7	4756.2	20784	115060	143.95	1.37486	45.34	1.2	7.7	26.0	24.8
10	2–Methylpentane	4395.8	4748.1	20757	113852	138.67	1.37145	45.34	1.2	(7.7)	73.5	73.4
11	3–Methylpentane	4398.7	4751.0	20768	115823	140.09	1.37652	45.34	(1.2)	(7.7)	74.3	74.5
12	Neohexane	4382.6	4735.0	20710	112932	131.24	1.36876	45.34	1.2	(7.7)	93.4	91.8
13	2,3–Dimethylbutane	4391.7	4744.0	20742	115243	136.08	1.37495	45.34	(1.2)	(7.7)	94.3	+0.3[f]
14	n–Heptane	5100.2	5502.8	20681	118668	136.01	1.38764	52.50	1.0	7.0	0.0	0.0
15	2–Methylhexane	5092.1	5494.8	20658	117627	131.59	1.38485	52.50	(1.0)	(7.0)	46.4	42.4
16	3–Methylhexane	5095.2	5497.8	20668	119192	132.11	1.38864	52.50	(1.0)	(7.0)	55.8	52.0
17	3–Ethylpentane	5098.2	5500.9	20679	121158	132.83	1.39339	52.50	(1.0)	(7.0)	69.3	65.0
18	2,2–Dimethylpentane	5079.4	5482.1	20620	116585	125.13	1.38215	52.50	(1.0)	(7.0)	95.6	92.8
19	2,4–Dimethylpentane	5084.3	5487.0	20636	116531	126.58	1.38145	52.50	(1.0)	(7.0)	83.8	83.1
20	3,3–Dimethylpentane	5085.0	5487.6	20638	120031	127.21	1.39092	52.50	(1.0)	(7.0)	86.6	80.8
21	Triptane	5081.0	5483.6	20627	119451	124.21	1.38944	52.50	(1.0)	(7.0)	+0.1[f]	+1.8[f]
22	n–Octane	5796.7	6249.7	20604	121419	129.53	1.39743	59.65	0.96	—	—	—
23	Diisobutyl	5781.3	6234.3	20564	119662	122.8	1.39246	59.65	(0.98)	—	55.7	55.2
24	Isooctane	5779.8	6232.8	20570	119388	116.71	1.39145	59.65	1.0	—	100.	100.
25	n–Nonane	6493.3	6996.5	20544	123613	123.76	1.40542	66.81	0.87[s]	2.9	—	—
26	n–Decane	7188.6	7742.1	20494	125444	118.68	1.41189	73.97	0.78[s]	2.6	—	—
27	Cyclopentane	3512.0	3763.7	20188	126296	167.34	1.40645	35.79	(1.4)	—	84.9[j]	+0.1[f]
28	Methylcyclopentane	4198.4	4500.4	20130	126477	147.83	1.40970	42.95	(1.2)	8.35	80.0	91.3
29	Cyclohexane	4178.8	4480.8	20035	130849	153.0	1.42623	42.95	1.3	7.8	77.2	83.0
30	Methylcyclohexane	4862.8	5215.2	20001	129066	136.3	1.42312	50.11	1.2	—	71.1	74.8
31	Ethylene	1499.0	1599.0	—	—	207.57	—	14.32	2.7	34.0	75.6	+0.3[f]
32	Propene	2182.7	2333.7	—	—	188.18	—	21.48	2.0	10.0	84.9	+0.2[f]
33	1–Butene	2879.4	3080.7	20678	103659	167.94	—	28.63	1.6	9.3	80.8[j]	97.4
34	Cis–2–Butene	2871.7	3073.1	20611	107754	178.91	—	28.63	(1.6)	—	83.5	100.
35	Trans–2–Butene	2866.8	3068.2	20584	104690	174.39	—	28.63	(1.6)	—	—	—
36	Isobutene	2860.4	3061.8	20548	102863	169.48	—	28.63	(1.6)	—	—	—
37	1–Pentene	3575.2	3826.9	20548	110610	154.46	1.37148	35.79	1.4	8.7	77.1	90.9
38	1,2–Butadiene	2789.0	2940.0	20447	112172	(181.)	—	26.25	(2.0)	(12.)	—	—
39	1,3–Butadiene	2730.0	2881.0	20047	104826	(174.)	—	26.25	2.0	11.5	—	—
40	Isoprene	3410.8	3612.1	19964	114194	(153.)	1.42194	33.41	(1.5)	—	81.0	99.1
41	Acetylene	1422.4	1472.8	—	—	—	—	11.93	2.5	80.	—	—
42	Benzene	3590.7	3741.7	17992	132655	169.31	1.50112	35.79	1.3[g]	7.9[g]	+2.8[f]	—
43	Toluene	4273.3	4474.7	18252	132656	154.84	1.49693	42.95	1.2[g]	7.1[g]	+0.3[f]	+5.8[f]
44	Ethylbenzene	4970.0	5221.7	18494	134414	144.0	1.49588	50.11	0.99[g]	6.7[g]	97.9	+0.8[f]
45	o–Xylene	4958.3	5210.0	18445	136069	149.1	1.50545	50.11	1.1[g]	6.4[g]	100.	—
46	m–Xylene	4956.8	5208.5	18441	133568	147.2	1.49722	50.11	1.1[g]	6.4[g]	+2.8[f]	+4.0[f]
47	p–Xylene	4956.9	5208.5	18445	133136	144.52	1.49582	50.11	1.1[g]	6.6[g]	+1.2[f]	+3.4[f]
48	Styrene	4828.7	5030.0	18150	137849	(151.)	1.54682	47.72	1.1	6.1	+0.2[f]	>+3.[f]
49	Isopropylbenzene	5661.4	5963.4	18665	134317	134.3	1.49145	57.27	0.88[g]	6.5[g]	99.3	+2.1[f]
50	Methyl Alcohol	—	—	9760	64771	473.(2)	1.3288(8)	7.16	6.72(5)	36.50	—	—
51	Ethyl Alcohol	—	—	12780	84600	367.(2)	1.3614(8)	14.32	3.28(5)	18.95	—	—
52	Carbon Monoxide	—	321.(13)	—	—	92.7(14)	—	2.39	12.50(5)	74.20	—	—
53	Carbon Dioxide	—	—	—	—	238.2[n](14)	—	—	—	—	—	—
54	Hydrogen Sulfide	588.(16)	637.(16)	—	—	235.6(7)	—	7.16	4.30(5)	45.50	—	—
55	Sulfur Dioxide	—	—	—	—	166.7(14)	—	—	—	—	—	—
56	Ammonia	359.(16)	434.(16)	—	—	587.2(14)	—	3.58	15.50(5)	27.00	—	—
57	Air	—	—	—	—	92.(3)	—	—	—	—	—	—
58	Hydrogen	274.(13)	324.(13)	—	—	193.9(14)	—	2.39	4.00(5)	74.20	—	—
59	Oxygen	—	—	—	—	91.6(14)	—	—	—	—	—	—
60	Nitrogen	—	—	—	—	87.8(14)	—	—	—	—	—	—
61	Chlorine	—	—	—	—	123.8(14)	—	—	—	—	—	—
62	Water	—	—	—	—	970.3(12)	1.3330(8)	—	—	—	—	—
63	Helium	—	—	—	—	—	—	—	—	—	—	—
64	Hydrogen Chloride	—	—	—	—	185.5(14)	—	—	—	—	—	—

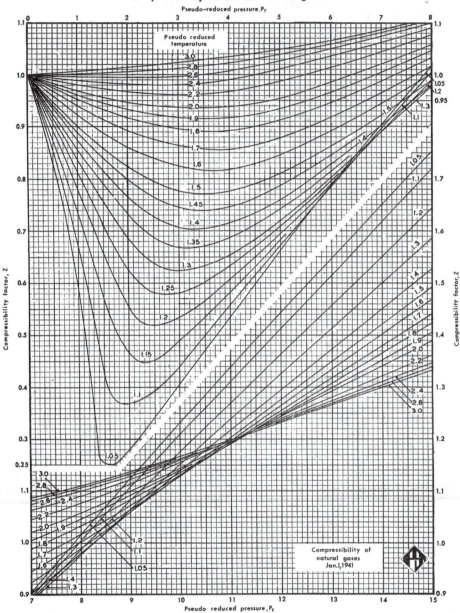

Compressibility factors for natural gas

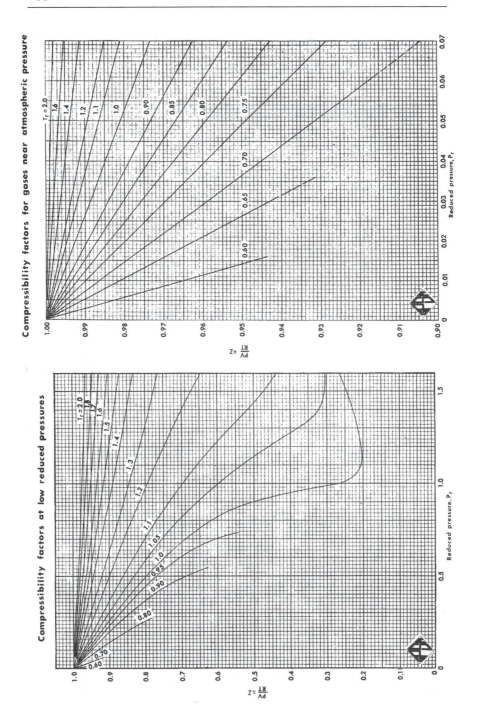

Compressibility factors for gases near atmospheric pressure

Compressibility factors at low reduced pressures

Compressibility of low-molecular-mass natural gases

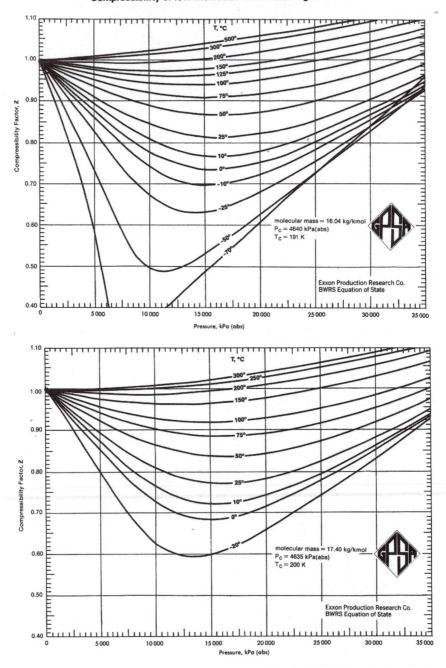

Compressibility of low-molecular-mass natural gases

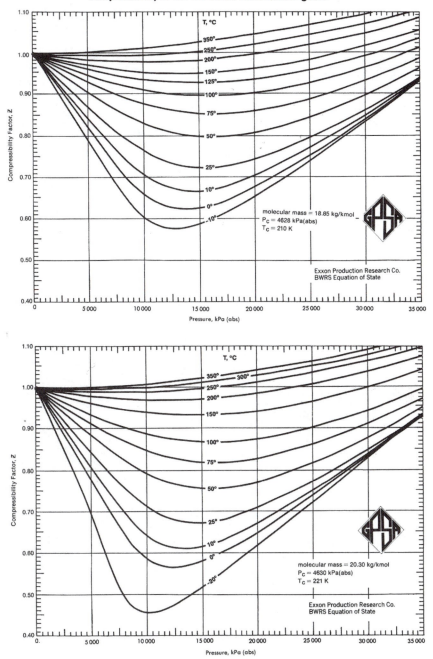

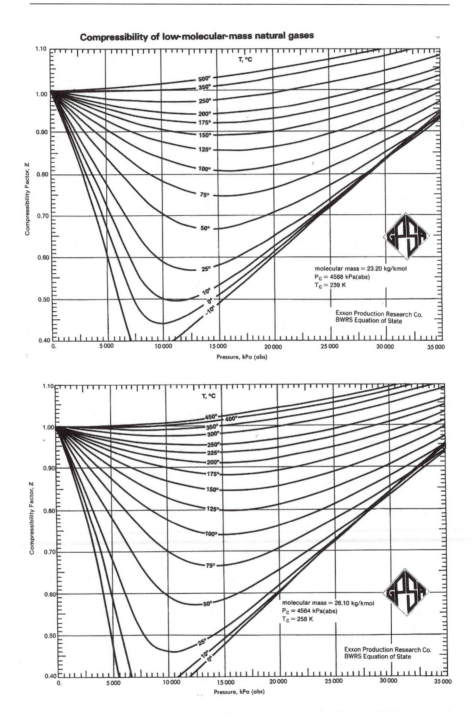

PHYSICAL CONSTANTS OF HYDROCARBONS(27)

No.	Compound	Formula	Molecular mass	Boiling point, °C 101.3250 kPa (abs) [1.]	Vapor pressure, kPa (abs) 40°C [2.]	Freezing point, °C 101.3250 kPa (abs) [3.]	Critical constants Pressure, kPa (abs)	Temperature, K	Volume, m³/kg
1	Methane	CH_4	16.043	−161.52(28)	(35 000.)	−182.47^d	4 604.	190.55	0.006 17
2	Ethane	C_2H_6	30.070	−88.58	(6 000.)	−182.80^d	4 880.	305.43	0.004 92
3	Propane	C_3H_8	44.097	−42.07	1 341.	−187.68^d	4 249.	369.82	0.004 60
4	n-Butane	C_4H_{10}	58.124	−0.49	377.	−138.36	3 797.	425.16	0.004 39
5	Isobutane	C_4H_{10}	58.124	−11.81	528.	−159.60	3 648.	408.13	0.004 52
6	n-Pentane	C_5H_{12}	72.151	36.06	115.66	−129.73	3 369.	469.6	0.004 21
7	Isopentane	C_5H_{12}	72.151	27.84	151.3	−159.90	3 381.	460.39	0.004 24
8	Neopentane	C_5H_{12}	72.151	9.50	269.	−16.55	3 199.	433.75	0.004 20
9	n-Hexane	C_6H_{14}	86.178	68.74	37.28	−95.32	3 012.	507.4	0.004 29
10	2-Methylpentane	C_6H_{14}	86.178	60.26	50.68	−153.66	3 010.	497.45	0.004 26
11	3-Methylpentane	C_6H_{14}	86.178	63.27	45.73	—	3 124.	504.4	0.004 26
12	Neohexane	C_6H_{14}	86.178	49.73	73.41	−99.870	3 081.	488.73	0.004 17
13	2,3-Dimethylbutane	C_6H_{14}	86.178	57.98	55.34	−128.54	3 127.	499.93	0.004 15
14	n-Heptane	C_7H_{16}	100.205	98.42	12.34	−90.582	2 736.	540.2	0.004 31
15	2-Methylhexane	C_7H_{16}	100.205	90.05	17.22	−118.27	2 734.	530.31	0.004 20
16	3-Methylhexane	C_7H_{16}	100.205	91.85	16.16	—	2 814.	535.19	0.004 03
17	3-Ethylpentane	C_7H_{16}	100.205	93.48	15.27	−118.60	2 891.	540.57	0.004 15
18	2,2-Dimethylpentane	C_7H_{16}	100.205	79.19	26.32	−123.81	2 773.	520.44	0.004 15
19	2,4-Dimethylpentane	C_7H_{16}	100.205	80.49	24.84	−119.24	2 737.	519.73	0.004 17
20	3,3-Dimethylpentane	C_7H_{16}	100.205	86.06	20.93	−134.46	2 945.	536.34	0.004 13
21	Triptane	C_7H_{16}	100.205	80.88	25.40	−24.91	2 954.	531.11	0.003 97
22	n-Octane	C_8H_{18}	114.232	125.67	4.143	−56.76	2 486.	568.76	0.004 31
23	Diisobutyl	C_8H_{18}	114.232	109.11	8.417	−91.200	2 486.	549.99	0.004 22
24	Isooctane	C_8H_{18}	114.232	99.24	12.96	−107.38	2 568.	543.89	0.004 10
25	n-Nonane	C_9H_{20}	128.259	150.82	1.40	−53.49	2 288.	594.56	0.004 27
26	n-Decane	$C_{10}H_{22}$	142.286	174.16	0.4732	−29.64	2 099.	617.4	0.004 24
27	Cyclopentane	C_5H_{10}	70.135	49.25	73.97	−93.866	4 502.	511.6	0.003 71
28	Methylcyclopentane	C_6H_{12}	84.162	71.81	33.85	−142.46	3 785.	532.73	0.003 79
29	Cyclohexane	C_6H_{12}	84.162	80.73	24.63	6.554	4 074.	553.5	0.003 68
30	Methylcyclohexane	C_7H_{14}	98.189	100.93	12.213	−126.59	3 472.	572.12	0.003 75
31	Ethene (Ethylene)	C_2H_4	28.054	−103.77(29)	—	−169.15^d	5 041.	282.35	0.004 67
32	Propene (Propylene)	C_3H_6	42.081	−47.72	1 596.	−185.25^d	4 600.	364.85	0.004 30
33	1-Butene (Butylene)	C_4H_8	56.108	−6.23	451.9	−185.35^d	4 023.	419.53	0.004 28
34	cis-2-Butene	C_4H_8	56.108	3.72	337.6	−138.91	4 220.	435.58	0.004 17
35	trans-2-Butene	C_4H_8	56.108	0.88	365.8	−105.55	4 047.	428.63	0.004 24
36	Isobutene	C_4H_8	56.108	−6.91	452.3	−140.35	3 999.	417.90	0.004 26
37	1-Pentene	C_5H_{10}	70.135	29.96	141.65	−165.22	3 529.	464.78	0.004 22
38	1,2-Butadiene	C_4H_6	54.092	10.85	269.	−136.19	(4 502.)	(444.)	(0.004 05)
39	1,3-Butadiene	C_4H_6	54.092	−4.41	434.	−108.91	4 330.	425.	0.004 09
40	Isoprene	C_5H_8	68.119	34.07	123.77	−145.95	(3 850.)	(484.)	(0.004 06)
41	Acetylene	C_2H_2	26.038	−84.88^e	—	−80.8^d	6 139.	308.33	0.004 34
42	Benzene	C_6H_6	78.114	80.09	24.38	5.533	4 898.	562.16	0.003 28
43	Toluene	C_7H_8	92.141	110.63	7.895	−94.991	4 106.	591.80	0.003 43
44	Ethylbenzene	C_8H_{10}	106.168	136.20	2.87	−94.975	3 609.	617.20	0.003 43
45	o-Xylene	C_8H_{10}	106.168	144.43	2.05	−25.18	3 734.	630.33	0.003 48
46	m-Xylene	C_8H_{10}	106.168	139.12	2.53	−47.87	3 536.	617.05	0.003 54
47	p-Xylene	C_8H_{10}	106.168	138.36	2.65	13.26	3 511.	616.23	0.003 56
48	Styrene	C_8H_8	104.152	145.14	1.85	−30.61	3 999.	647.6	0.003 38
49	Isopropylbenzene	C_9H_{12}	120.195	152.41	1.47	−96.035	3 209.	631.1	0.003 57
50	Methyl alcohol	CH_4O	32.042	64.54	35.43	−97.68	8 096.	512.64	0.003 68
51	Ethyl alcohol	C_2H_6O	46.069	78.29	17.70	−114.1	6 383.	513.92	0.003 62
52	Carbon monoxide	CO	28.010	−191.49	—	−205.0^d	3 499.(33)	132.92(33)	0.003 32(33)
53	Carbon dioxide	CO_2	44.010	−78.51^e	—	−56.57^d	7 382.(33)	304.19(33)	0.002 14(33)
54	Hydrogen sulfide	H_2S	34.076	−60.31	2 881.	−85.53^d	9 005.	373.5	0.002 87
55	Sulfur dioxide	SO_2	64.059	−10.02	630.8	−75.48^d	7 894.	430.8	0.001 90
56	Ammonia	NH_3	17.031	−33.33(30)	1 513.	−77.74^d	11 280.	405.6	0.004 25
57	Air	$N_2 + O_2$	28.964	−194.2(2)	—	—	3 771.(2)	132.4(2)	0.003 23(3)
58	Hydrogen	H_2	2.016	−252.87^v	—	−259.2^d	1 297.	33.2	0.032 24
59	Oxygen	O_2	31.999	−182.962^v	—	−218.8^d	5 081.	154.7(33)	0.002 29
60	Nitrogen	N_2	28.013	−195.80(31)	—	−210.0^d	3 399.	126.1	0.003 22
61	Chlorine	Cl_2	70.906	−34.03	1 134.	−101.0^d	7 711.	417.	0.001 75
62	Water	H_2O	18.015	100.00^v	7.377	0.00	22 118.	647.3	0.003 18
63	Helium	He	4.003	−268.93(32)	—	—	227.5(32)	5.2(32)	0.014 36(32)
64	Hydrogen chloride	HCl	36.461	−85.00	6 304.	−114.18^d	8 309.	324.7	0.002 22

PHYSICAL CONSTANTS OF HYDROCARBONS(27)

Relative density 15 °C/15 °C a,b	kg/m³ *,a (mass in vacuum)	kg/m³ *,a,c (Apparent mass in air)	m³/kmol	Temperature coefficient of density, at 15 °C *,a −1/°C	Pitzer acentric factor, ω	Compressibility factor of real gas, Z 101.3250 kPa (abs), 15 °C	Relative density Air = 1	Specific volume m³/kg	Volume ratio gas/(liquid in vacuum)	Cp Ideal gas	Cp Liquid	No.
(0.3)^i	(300.)^i	(300.)^i	(0.05)^i	—	0.0126	0.9981	0.5539	1.474	(442.)^i	2.204	—	1
0.3581^h,x	357.8^h,x	356.6^h	0.084 04^h	—	0.0978	0.9915	1.0382	0.7863	281.3^h	1.706	3.807	2
0.5083^h	507.8^h,x	506.7^h	0.086 84^h	0.002 74^h	0.1541	0.9810	1.5225	0.5362	272.3^h	1.625	2.476	3
0.5847^h	584.2^h	583.1^h	0.099 49^h	0.002 11^h	0.2015	0.9641	2.0068	0.4068	237.6^h	1.652	2.366(41)	4
0.5637^h	563.2^h	562.1^h	0.103 2^h	0.002 14^h	0.1840	0.9665	2.0068	0.4068	229.1^h	1.616	2.366(41)	5
0.6316	631.0	629.9	0.114 3	0.001 57	0.2524	0.942^t	2.4911	0.3277	206.8	1.622	2.292(41)	6
0.6250	624.4	623.3	0.115 6	0.001 62	0.2286	0.948^t	2.4911	0.3277	204.6	1.600	2.239	7
0.5972^h	596.7^h	595.6^h	0.120 9^h	0.001 87^h	0.1967	0.9538	2.4911	0.3277	195.5^h	1.624	2.317	8
0.6644	663.8	662.7	0.129 8	0.001 35	0.2998	0.910^t	2.9753	0.2744	182.1	1.613	2.231	9
0.6583	657.7	656.6	0.131 0	0.001 40	0.2784	—	2.9753	0.2744	180.5	1.602	2.205	10
0.6694	668.8	667.7	0.128 9	0.001 35	0.2741	—	2.9753	0.2744	183.5	1.578	2.170	11
0.6545	653.9	652.8	0.131 8	0.001 40	0.2333	—	2.9753	0.2744	179.4	1.593	2.148	12
0.6668	666.2	665.1	0.129 4	0.001 35	0.2475	—	2.9753	0.2744	182.8	1.566	2.146	13
0.6886	688.0	686.9	0.145 6	0.001 24	0.3494	0.852^t	3.4596	0.2360	162.4	1.606	2.209	14
0.6835	682.8	681.7	0.146 8	0.001 22	0.3303	—	3.4596	0.2360	161.1	1.595	2.183	15
0.6921	691.5	690.4	0.144 9	0.001 24	0.3239	—	3.4596	0.2360	163.2	1.584	2.137	16
0.7032	702.6	701.5	0.142 6	0.001 26	0.3107	—	3.4596	0.2360	165.8	1.613	2.150	17
0.6787	678.0	676.9	0.147 8	0.001 30	0.2876	—	3.4596	0.2360	160.0	1.613	2.161	18
0.6777	677.1	676.0	0.148 0	0.001 30	0.3031	—	3.4596	0.2360	159.8	1.651	2.193	19
0.6980	697.4	696.3	0.143 7	0.001 17	0.2681	—	3.4596	0.2360	164.6	1.603	2.099	20
0.6950	694.4	693.3	0.144 3	0.001 24	0.2509	—	3.4596	0.2360	163.9	1.578	2.088	21
0.7073	706.7	705.6	0.161 6	0.001 12	0.3981	0.783^t	3.9439	0.2070	146.3	1.601	2.191	22
0.6984	697.7	696.6	0.163 7	0.001 17	0.3564	—	3.9439	0.2070	144.4	1.573	2.138	23
0.6966	696.0	694.9	0.164 1	0.001 17	0.3041	—	3.9439	0.2070	144.1	1.599	2.049	24
0.7224	721.7	720.6	0.177 7	0.001 13	0.4452	—	4.4282	0.1843	133.0	1.598	2.184	25
0.7346	733.9	732.8	0.193 9	0.000 99	0.4904	—	4.9125	0.1662	122.0	1.595	2.179	26
0.7508	750.2	749.1	0.093 49	0.001 26	0.1945	0.949^t	2.4215	0.3371	252.9	1.133	1.763	27
0.7541	754.4	752.3	0.111 7	0.001 28	0.2308	—	2.9057	0.2809	211.7	1.258	1.843	28
0.7838	783.1	782.0	0.107 5	0.001 22	0.2098	—	2.9057	0.2809	220.0	1.211	1.811	29
0.7744	773.7	772.6	0.126 9	0.001 13	0.2364	—	3.3900	0.2408	186.3	1.324	1.839	30
—	—	—	—	—	0.0869	0.9938	0.9686	0.8428	—	1.514	—	31
0.5231^h	522.6^h,x	521.5^h	0.080 69^h	0.003 40^h	0.1443	0.9844	1.4529	0.5619	293.6^h	1.480	2.443	32
0.6019^h	601.4^h	600.3^h	0.093 30^h	0.002 09^h	0.1949	0.9703	1.9372	0.4214	253.4^h	1.483	2.237	33
0.6277^h	627.1^h	626.0^h	0.089 47^h	0.001 76^h	0.2033	0.9660	1.9372	0.4214	264.3^h	1.366	2.241(42)	34
0.6105^h	610.0^h	608.9^h	0.091 98^h	0.001 93^h	0.2126	0.9661	1.9372	0.4214	257.1^h	1.528	2.238	35
0.6010^h	600.5^h	599.4^h	0.093 44^h	0.002 16^h	0.2026	0.9688	1.9372	0.4214	253.1^h	1.547	2.296	36
0.6462	645.6	644.5	0.108 6	0.001 60	0.2334	0.948^t	2.4215	0.3371	217.7	1.519	2.241(43)	37
0.6576^h	657.^h	656.^h	0.082.33^h	0.001 76^h	(0.2540)	(0.969)	1.8676	0.4371	287.2^h	1.446	2.262	38
0.6280^h	627.4^h	626.3^h	0.086 22^h	0.002 03^h	0.1971	(0.965)	1.8676	0.4371	274.2^h	1.426	2.124	39
0.6866	686.0	684.9	0.099 30	0.001 55	(0.1567)	0.949^t	2.3519	0.3471	238.1	1.492	2.171	40
0.615^h	—	—	—	—	0.1893	0.9925	0.8990	0.9081	—	1.659	—	41
0.8850	884.2	883.1	0.088 34	0.001 19	0.2095	0.929^t	2.6989	0.3027	267.6	1.014	1.715	42
0.8723	871.6	870.5	0.105 7	0.001 08	0.2633	0.903^t	3.1812	0.2566	223.7	1.085	1.677	43
0.8721	871.3	870.5	0.121 9	0.000 97	0.3031	—	3.6655	0.2227	194.0	1.168	1.721	44
0.8850	884.2	883.1	0.120 1	0.000 99	0.3113	—	3.6655	0.2227	196.9	1.218	1.741	45
0.8691	868.3	867.2	0.122 3	0.000 97	0.3257	—	3.6655	0.2227	193.4	1.163	1.696	46
0.8661	865.3	864.2	0.122 7	0.000 97	0.3214	—	3.6655	0.2227	192.7	1.157	1.708	47
0.9115	910.6	909.5	0.114 4	0.001 03	0.1997	—	3.5959	0.2270	206.7	1.133	1.724	48
0.8667	866.0	864.9	0.139 0	0.000 97	0.3260	—	3.4198	0.1967	170.4	1.219	1.732	49
0.7967	796.0	794.9	0.040 25	0.001 17	0.5648	—	1.1063	0.7379	587.4	1.352	2.484	50
0.7922	791.5	790.4	0.058 20	0.001 07	0.6608	—	1.5906	0.5132	406.2	1.389	2.348	51
0.7893^m	788.6^m(34)	—	0.035 52^m	—	0.0442	0.9995	0.9671	0.8441	—	1.040	—	52
0.8226^h	821.9^h(35)	820.8^h	0.053 55^h	—	0.2667	0.9943	1.5195	0.5373	441.6^h	0.8330	—	53
0.7897^h	789.0^h,x(36)	787.9^h	0.043 19^h	—	0.0920	0.9903	1.1765	0.6939	547.5^h	0.9960	2.08(36)	54
1.397^h	1396.^h,x(36)	1395.^h	0.045 89^h	—	0.2548	0.9801^t	2.2117	0.3691	515.3^h	0.6062	1.359(36)	55
0.6183^h	617.7^h,x(30)	616.6^h	0.027 57^h	—	0.2576	0.9899(30)	0.5880	1.388	857.4	2.079	4.693(30)	56
0.856^m(36)	855.^m	—	0.033 9^m	—	—	0.9996	1.0000	0.8163	—	1.005	—	57
0.07106^m	71.00^m(37)	—	0.028 39^m	—	−0.219^w	1.0006	0.0696	11.73	—	14.24	—	58
1.1420^m(25)	1141.^m(38)	—	0.028 04^m	—	0.0200	0.9993(39)	1.1048	0.7389	—	0.9166	—	59
0.8093^m(28)	808.6^m(31)	—	0.034 64^m	—	0.0372	0.9997	0.9672	0.8441	—	1.040	—	60
1.426	1424.5	1423.5	0.049 78	—	0.0737	(0.9875)^t(36)	2.4481	0.3335	475.0	0.4760	—	61
1.000	999.1	998.0	0.018 03	0.000 14	0.3434	—	0.6220	1.312	1311.	1.862	4.191	62
0.1251^m	125.0^m(32)	—	0.032 02^m	—	0	1.000 5(40)	0.1382	5.907	—	5.192	—	63
0.8538	853.0^x	851.9	0.042 74	0.006 03	0.1232	—	1.2588	0.6485	553.2	0.7991	—	64

PHYSICAL CONSTANTS OF HYDROCARBONS(27)

No.	Compound	10. Heating value, 15°C*				11. Heat of vaporization, 101.3250 kPa (abs) at boiling point, kJ/kg	12. Refractive index, ^{a}, n_D 15°C	13. Air required for combustion, ideal gas* $m^3(air)/m^3(gas)$	Flammability limits, vol % in air mixture		ASTM octane number	
		Net	Gross ^{p}						Lower	Higher	Motor method D-357	Research method D-908
		MJ/m³ Ideal gas, 101.3250 kPa (abs)	MJ/m³ Ideal gas, 101.3250 kPa (abs)	MJ/kg Liquid (Mass in vacuum)	MJ/m³ Liquid (Mass in vacuum)							
1	Methane	33.936	37.694	—		509.86	—	9.54	5.0	15.0	—	—
2	Ethane	60.395	66.032	51.586^h	18 458^h	489.36	1.214 04^h	16.70	2.9	13.0	+.05^f	+1.6^{j,f}
3	Propane	86.456	93.972	50.008^h	25 394^h	425.73	1.219 05^h	23.86	2.1	9.5	97.1	+1.9^{j,f}
4	n-Butane	112.384	121.779	49.158^h	28 718^h	385.26	1.332 92^h	31.02	1.8	8.4	89.6^g	93.8^j
5	Isobutane	112.031	121.426	49.044^h	27 621^h	366.40		31.02	1.8	8.4	97.6	+0.10^{j,f}
6	n-Pentane	138.380	149.654	48.667	30 709	357.22	1.360 24	38.18	1.4	8.3	62.6^j	61.7^j
7	Isopentane	138.044	149.319	48.579	30 333	342.20	1.356 58	38.18	1.4	(8.3)	90.3	92.3
8	Neopentane	137.465	148.739	48.427^h	28 896^h	315.34	1.345	38.18	1.4	(8.3)	80.2	85.5
9	n-Hexane	164.402	177.556	48.344	32 091	334.81	1.377 46	45.34	1.2	7.7	26.0	24.8
10	2-Methylpentane	164.075	177.229	48.273	31 749	322.52	1.374 17	45.34	1.2	(7.7)	73.5	73.4
11	3-Methylpentane	164.188	177.341	48.300	32 303	325.82	1.379 18	45.34	(1.2)	(7.7)	74.3	74.5
12	Neohexane	163.683	176.836	48.191	31 512	305.24	1.371 57	45.34	1.2	(7.7)	93.4	91.8
13	2,3-Dimethylbutane	164.025	177.179	48.269	32 157	316.50	1.377 59	45.34	(1.2)	(7.7)	94.3	+0.3^f
14	n-Heptane	190.398	205.431	48.104	33 095	316.33	1.390 17	52.50	1.0	7.0	0.0	0.0
15	2-Methylhexane	190.099	205.132	48.051	32 809	306.06	1.387 43	52.50	(1.0)	(7.0)	46.4	42.4
16	3-Methylhexane	190.243	205.276	48.082	33 249	307.27	1.391 19	52.50	(1.0)	(7.0)	55.8	52.0
17	3-Ethylpentane	190.327	205.359	48.101	33 796	308.94	1.395 94	52.50	(1.0)	(7.0)	69.3	65.0
18	2,2-Dimethylpentane	189.630	204.662	47.964	32 520	291.03	1.384 75	52.50	(1.0)	(7.0)	95.6	92.8
19	2,4-Dimethylpentane	189.803	204.836	48.000	32 501	294.41	1.384 08	52.50	(1.0)	(7.0)	83.8	83.1
20	3,3-Dimethylpentane	189.885	204.918	48.019	33 488	295.87	1.393 42	52.50	(1.0)	(7.0)	86.6	80.8
21	Triptane	189.690	204.722	47.982	33 319	288.90	1.391 96	52.50	(1.0)	(7.0)	+0.1^f	+1.8^f
22	n-Octane	216.374	233.286	47.919	33 865	301.26	1.399 81	59.65	0.96	—	—	—
23	Diisobutyl	215.797	232.709	47.832	33 372	285.69	1.394 88	59.65	(0.98)	—	55.7	55.2
24	Isooctane	215.732	232.644	47.843	33 299	271.44	1.393 92	59.65	1.0	—	100.	100.
25	n-Nonane	242.398	261.189	47.783	34 485	288.82	1.407 73	66.81	0.87^g	2.9	—	—
26	n-Decane	268.396	289.066	47.670	34 985	276.06	1.414 11	73.97	0.78^g	2.6	—	—
27	Cyclopentane	131.114	140.509	46.955	35 225	389.20	1.409 27	35.79	(1.4)	—	84.9^j	+0.1^f
28	Methylcyclopentane	156.757	168.032	46.825	35 278	345.51	1.412 40	42.95	(1.2)	8.35	80.0	91.3
29	Cyclohexane	156.034	167.308	46.606	36 497	355.95	1.428 92	42.95	1.3	7.8	77.2	83.0
30	Methylcyclohexane	181.567	194.720	46.525	35 997	317.03	1.425 66	50.11	1.2	—	71.1	74.8
31	Ethene (Ethylene)	55.942	59.700	—	—	482.77	—	14.32	2.7	34.0	75.6	+0.03^f
32	Propene (Propylene)	81.482	87.119	—	—	437.68	—	21.48	2.0	10.0	84.9	+0.2^f
33	1-Butene (Butylene)	107.475	114.991	48.081^h	28 916^h	390.60	—	28.63	1.6	9.3	80.8^j	97.4
34	cis-2-Butene	107.191	114.707	47.927^h	30 055^h	416.10	—	28.63	(1.6)	—	83.5	100.
35	trans-2-Butene	106.957	114.473	47.843^h	29 184^h	405.56	—	28.63	(1.6)	—	—	—
36	Isobutene	106.755	114.271	47.769^h	28 685^h	394.18	—	28.63	(1.6)	—	—	—
37	1-Pentene	133.465	142.860	47.788	30 852	359.25	1.374 61	35.79	1.4	8.7	77.1	90.9
38	1,2-Butadiene	104.118	109.755	47.504^h	31 210^h	(449.6)	—	26.25	(2.0)	(12.)	—	—
39	1,3-Butadiene	101.917	107.555	46.608^h	29 242^h	(418.7)	—	26.25	2.0	11.5	—	—
40	Isoprene	127.330	134.846	46.408	31 836	(385.2)	1.425 36	33.41	(1.5)	—	81.0	99.1
41	Acetylene	53.098	54.978	—	—	—	—	11.93	2.5	80.	—	—
42	Benzene	134.055	139.692	41.843	36 998	393.32	1.504 32	35.79	1.3^g	7.9^g	+2.8^f	—
43	Toluene	159.534	167.050	42.450	37 000	360.14	1.499 73	42.95	1.2^g	7.1^g	+0.3^f	+5.8^f
44	Ethylbenzene	185.555	194.950	43.014	37 478	334.98	1.498 56	50.11	0.99^g	6.7^g	97.9	+0.8^f
45	o-Xylene	185.092	194.487	42.900	37 935	346.80	1.507 95	50.11	1.1^g	6.4^g	100.	—
46	m-Xylene	185.020	194.415	42.891	37 245	342.47	1.499 80	50.11	1.1^g	6.4^g	+2.8^f	+4.0^f
47	p-Xylene	185.050	194.445	42.901	37 122	338.92	1.498 39	50.11	1.1^g	6.6^g	+1.2^f	+3.4^f
48	Styrene	180.290	187.806	42.213	38 439	(351.23)	1.549 69	47.72	1.1	6.1	+0.2^f	>+3.^{f}
49	Isopropylbenzene	211.328	222.603	43.410	37 591	312.25	1.494 00	57.27	0.88^g	6.5^g	99.3	+2.1^f
50	Methyl alcohol	28.601	32.360	22.685	18 057	1075.97	1.330 28	7.16	6.72(5)	36.50	—	—
51	Ethyl alcohol	54.062	59.699	29.707	23 513	840.54	1.363 45	14.32	3.28(5)	18.95	—	—
52	Carbon monoxide	11.959	11.959	—	—	215.70	1.000 36	2.39	12.50(5)	74.20	—	—
53	Carbon dioxide	0	0	—	—	573.27^n	1.000 49	—	—	—	—	—
54	Hydrogen sulfide	21.912	23.791	—	—	548.01	1.000 61	7.16	4.30(5)	45.50	—	—
55	Sulfur dioxide	—	—	—	—	387.74	1.000 62	—	—	—	—	—
56	Ammonia	17.301	20.121	—	—	1366.	1.000 36	3.58	15.50(5)	27.00	—	—
57	Air	—	—	—	—	214.	—	—	—	—	—	—
58	Hydrogen	10.230	12.091	—	—	450.4	1.000 13	2.39	4.00(5)	74.20	—	—
59	Oxygen	—	—	—	—	213.	1.000 27	—	—	—	—	—
60	Nitrogen	—	—	—	—	204.	1.000 28	—	—	—	—	—
61	Chlorine	—	—	—	—	288.0	1.387 8^y	—	—	—	—	—
62	Water	0	1.879	0	0	2257.	1.333 47	—	—	—	—	—
63	Helium	—	—	—	—		1.000 03	—	—	—	—	—
64	Hydrogen chloride	—	—	—	—	431.5	1.000 42	—	—	—	—	—

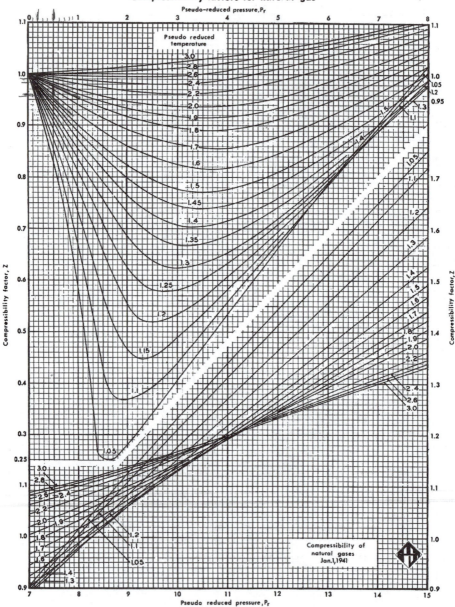

Compressibility factors for natural gas

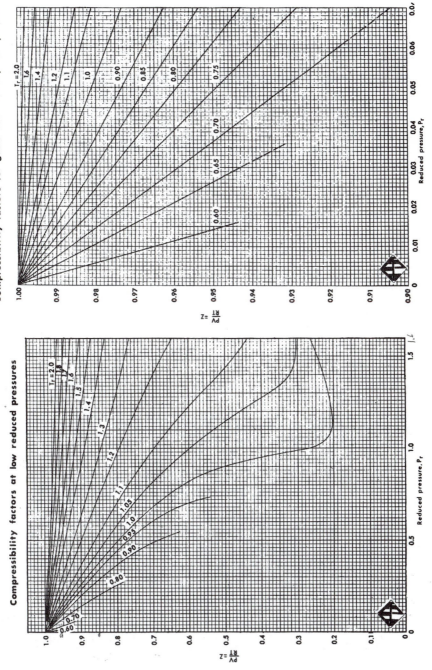

Compressibility factors for gases near atmospheric pressure

Compressibility factors at low reduced pressures

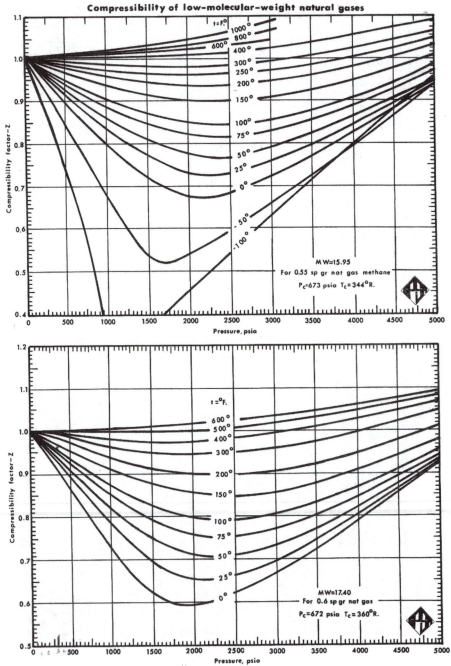

Compressibility of low-molecular-weight natural gases

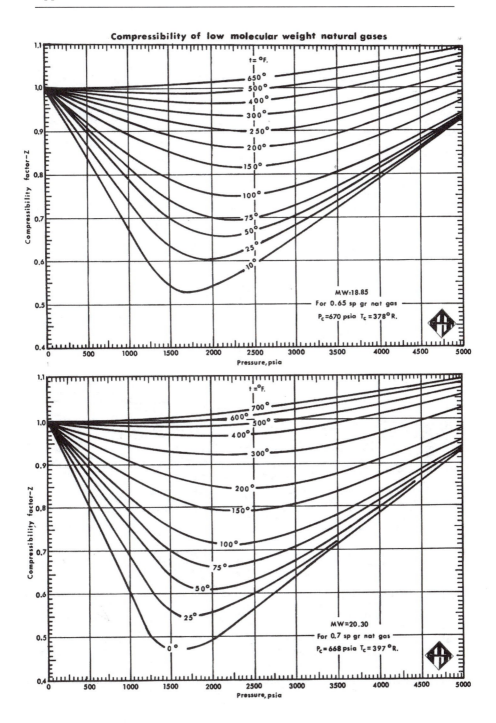

Compressibility of low molecular weight natural gases

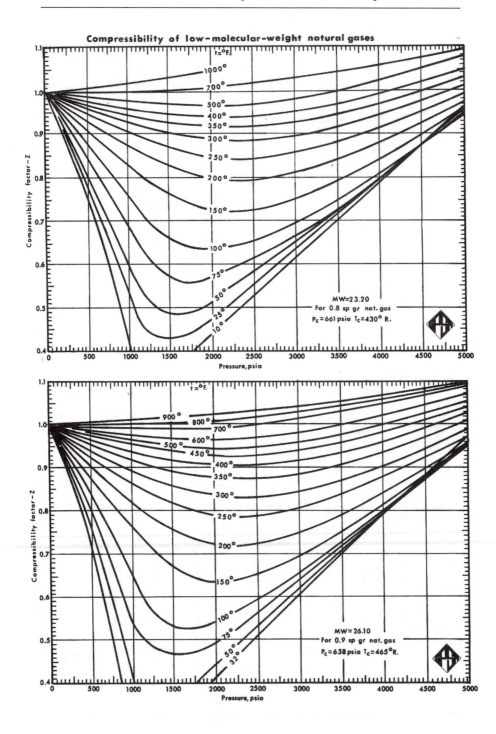

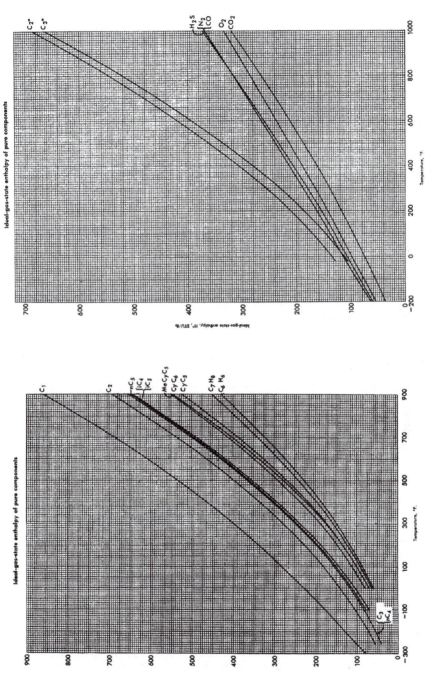

Ideal-gas-state enthalpy of pure components

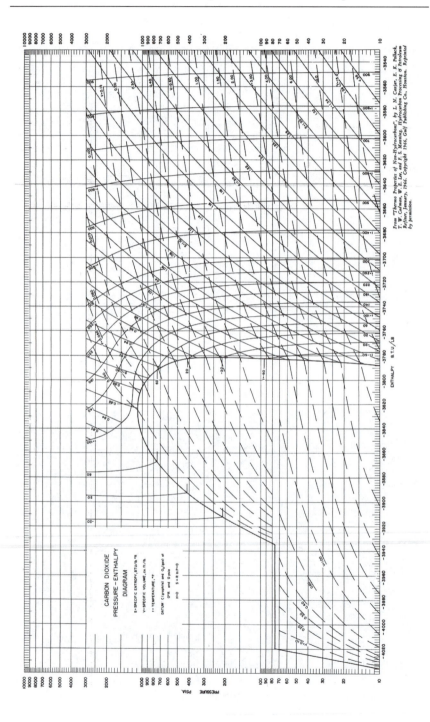

From "Thermo Properties of Non-Hydrocarbons", by L. N. Canjar, E. E. Pollock, T. W. Callinan, W. Lu, and F. S. Manning, Hydrocarbon Processing & Petroleum Refiner, January, 1966. Copyright 1966, Gulf Publishing Co., Houston. Reprinted by permission.

CARBON DIOXIDE
PRESSURE – ENTHALPY
DIAGRAM

S=SPECIFIC ENTROPY, BTU/lb.°R

V=SPECIFIC VOLUME, cu. ft./lb.

t=TEMPERATURE, °F

DATUM: C (graphite) and O₂(gas) at
0°R and 0 psia

H₂O   S = R ln P+O

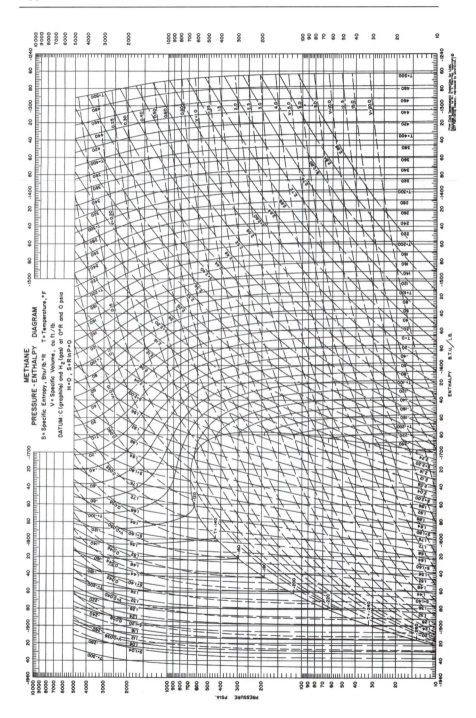

METHANE
PRESSURE-ENTHALPY DIAGRAM
S = Specific Entropy, Btu/lb.°R   T = Temperature, °F
V = Specific Volume, cu ft./lb.
DATUM: C (graphite) and H₂ (gas) at 0°R and 0 psia
H = 0,   S = R ln P = 0

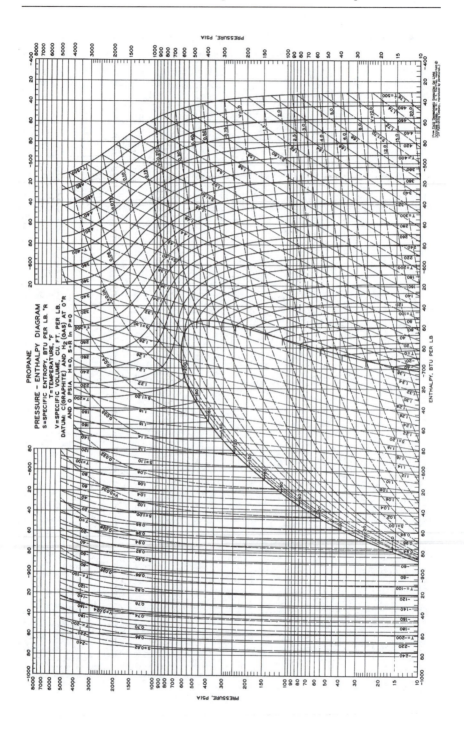

PROPANE

PRESSURE – ENTHALPY DIAGRAM

S=SPECIFIC ENTROPY, BTU PER LB. °R
T=TEMPERATURE, °F
V=SPECIFIC VOLUME, CU. FT. PER LB.
DATUM: C(GRAPHITE) AND H₂ (GAS) AT 0°R
AND 0 PSIA H=0, S+R ln P=0

PRESSURE, PSIA

ENTHALPY, BTU PER LB

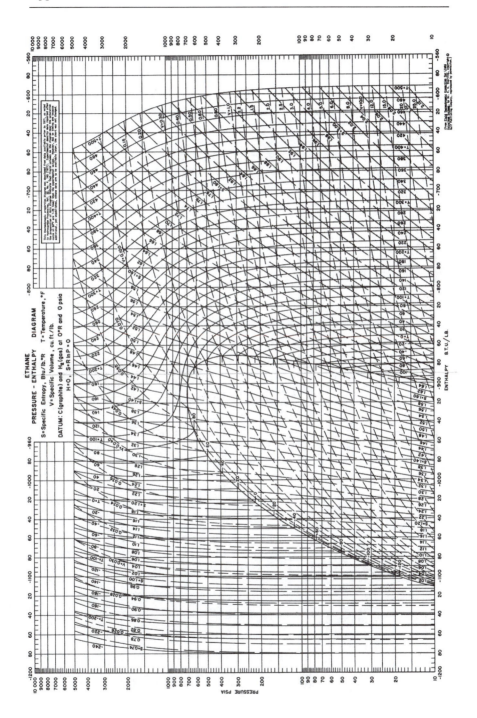

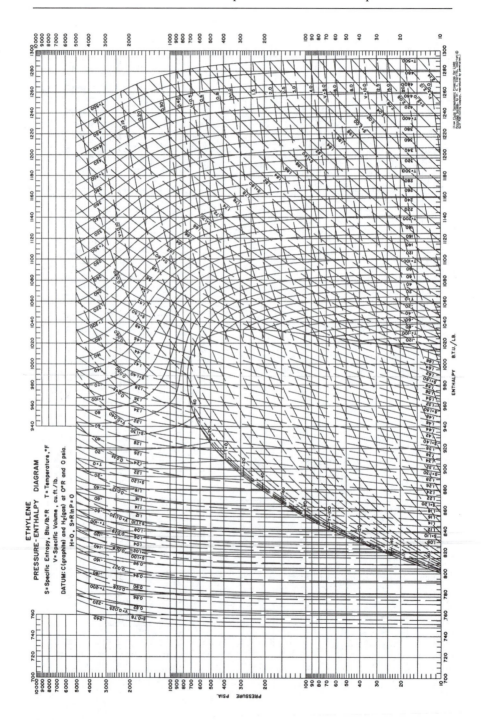

ETHYLENE
PRESSURE - ENTHALPY DIAGRAM

S = Specific Entropy, Btu/lb.°R   T = Temperature, °F
V = Specific Volume, cu. ft./lb.
DATUM: C(graphite) and H₂(gas) at 0°R and 0 psia.
H = 0, S + R ln P = 0

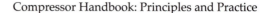

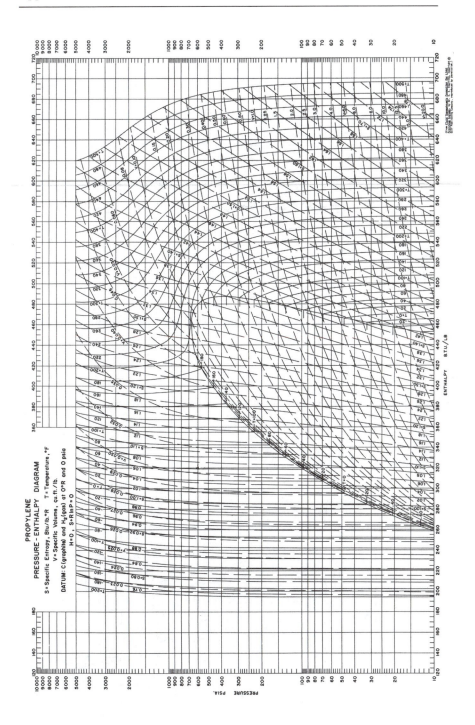

PROPYLENE
PRESSURE – ENTHALPY DIAGRAM

S = Specific Entropy, Btu/lb·°R   T = Temperature, °F

V = Specific Volume, cu.ft./lb.

DATUM: C (graphite) and H₂ (gas) at 0°R and 0 psia

H = 0, S = R ln P = 0

ENTHALPY B.T.U./LB.

PRESSURE PSIA.

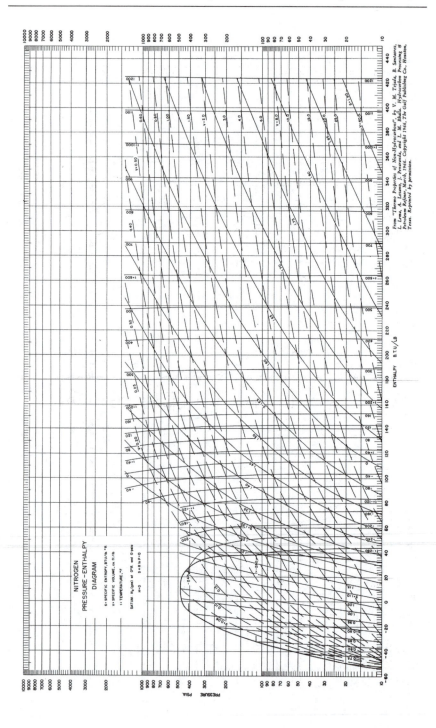

NITROGEN
PRESSURE-ENTHALPY
DIAGRAM

S-SPECIFIC ENTROPY, BTU/lb.·°R

V-SPECIFIC VOLUME, cu.ft./lb

T-TEMPERATURE,°F

DATUM: H₀ (gas) at 0°R, and 0 psia

H=0   S=0 at b.p·l·Q

PRESSURE PSIA

ENTHALPY   B.T.U./LB

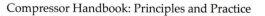

From "Thermo Properties of Non-Hydrocarbons", by V. M. Trijeda, B. Semisorro, L. Leau, A. Lozano, J. Penaranda, and L. M. Bhalla. Hydrocarbon Processing & Petroleum Refiner, March, 1966. Copyright 1966, The Gulf Publishing Co., Houston, Texas. Reprinted by permission.

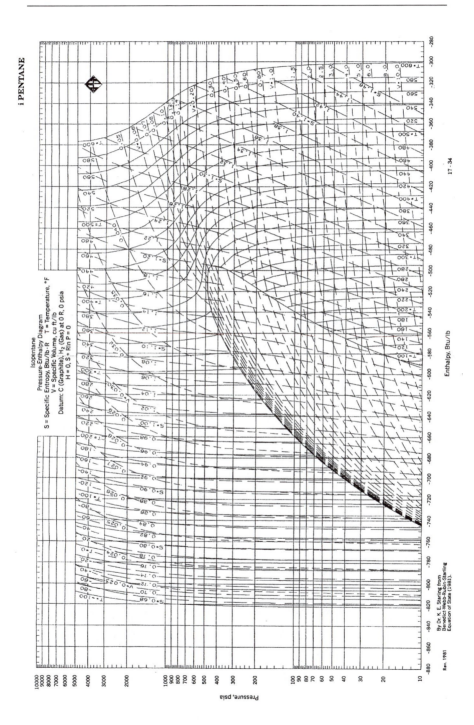

i PENTANE

Isopentane
Pressure-Enthalpy Diagram
S = Specific Entropy, Btu/lb-R   T = Temperature, °F
V = Specific Volume, cu ft/lb
Datum: C (Graphite), H₂ (Gas) at 0 R, 0 psia
H = 0, S + R ln P = 0

Pressure, psia

Enthalpy, Btu/lb

17 - 34

By Dr. K. E. Starling from
Benedict-Webb-Rubin-Starling
Equation of State (1981).

Rev. 1981

carbon dioxide

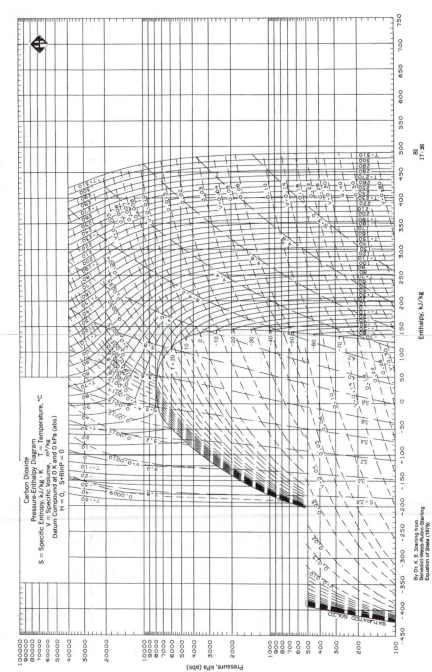

Carbon Dioxide
Pressure-Enthalpy Diagram
S = Specific Entropy, kJ/kg · K   T = Temperature, °C
V = Specific Volume,   m³/kg
Datum Compound at 0 K and 0 kPa (abs)
H = 0,   S+RlnP = 0

Enthalpy, kJ/kg

Pressure, kPa (abs)

By Dr. K. E. Starling from
Benedict-Webb-Rubin-Starling
Equation of State (1979)

SI
17 - 26

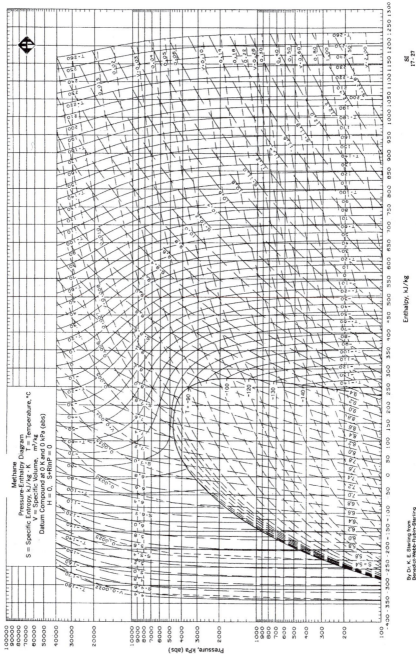

METHANE

Methane
Pressure-Enthalpy Diagram
S = Specific Entropy, kJ/kg · K   T = Temperature, °C
V = Specific Volume,   m³/kg
Datum Compound at 0 K and 0 kPa (abs)
H = 0, S+RlnP = 0

Enthalpy, kJ/kg

Pressure, kPa (abs)

By Dr. K. E. Starling from
Benedict-Webb-Rubin-Starling
Equation of State (1979).

SI
17-27

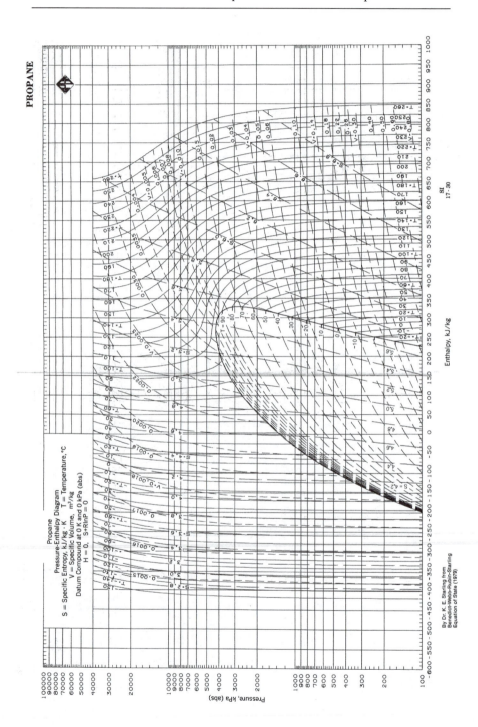

PROPANE

Propane
Pressure-Enthalpy Diagram
S = Specific Entropy, kJ/kg · K   T = Temperature, °C
V = Specific Volume, m³/kg
Datum Compound at 0 K and 0 kPa (abs)
H = 0,   S+RlnP = 0

Pressure, kPa (abs)

Enthalpy, kJ/kg

SI
17-30

By Dr. K. E. Starling from
Benedict-Webb-Rubin-Starling
Equation of State (1979)

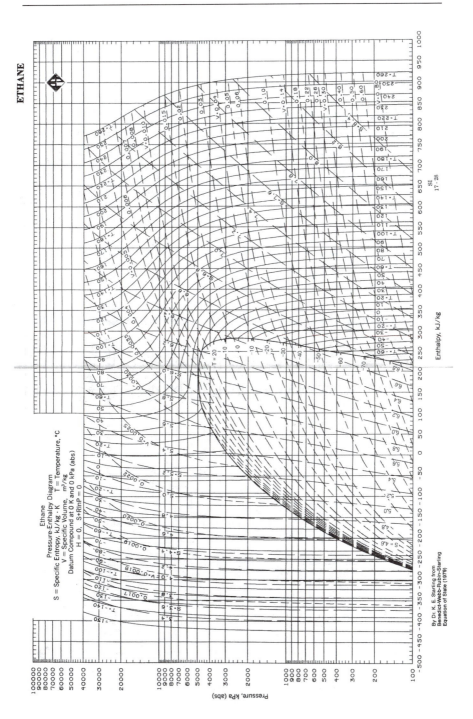

ETHANE

Ethane
Pressure-Enthalpy Diagram
S = Specific Entropy, kJ/kg · K   T = Temperature, °C
V = Specific Volume, m³/kg
Datum Compound at 0 K and 0 kPa (abs)
H = 0,   S+RlnP = 0

Pressure, kPa (abs)

Enthalpy, kJ/kg

SI
17 · 28

By Dr. K. E. Starling from
Benedict-Webb-Rubin-Starling
Equation of State (1979)

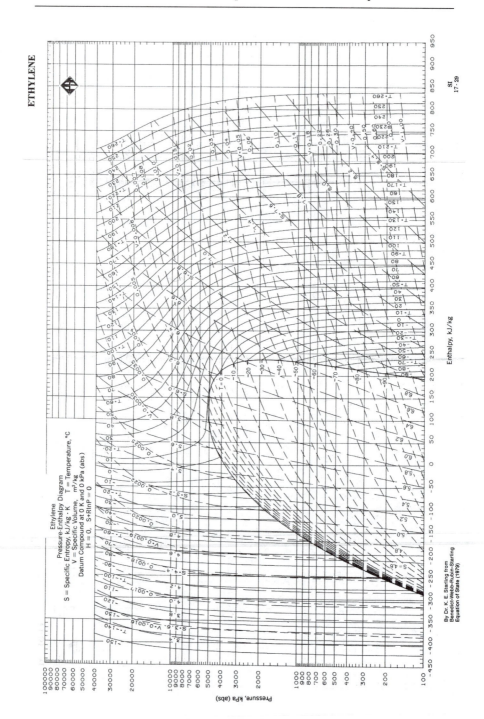

ETHYLENE

Ethylene
Pressure-Enthalpy Diagram
S = Specific Entropy, kJ/kg · K   T = Temperature, °C
V = Specific Volume,   m³/kg
Datum Compound at 0 K and 0 kPa (abs)
H = 0, S+RlnP = 0

Enthalpy, kJ/kg

Pressure, kPa (abs)

By Dr. K. E. Starling from
Benedict-Webb-Rubin-Starling
Equation of State (1979)

SI
17-29

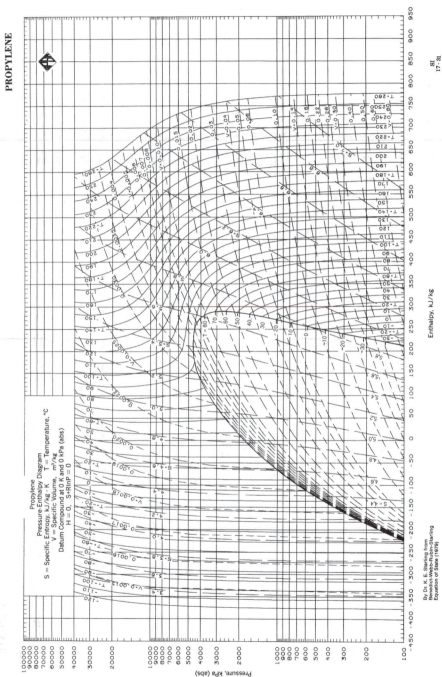

PROPYLENE

Propylene
Pressure-Enthalpy Diagram
S = Specific Entropy, kJ/kg · K   T = Temperature, °C
V = Specific Volume, m³/kg
Datum Compound at 0 K and 0 kPa (abs)
H = 0, S+RlnP = 0

Enthalpy, kJ/kg

Pressure, kPa (abs)

By Dr. K. E. Starling from
Benedict-Webb-Rubin-Starling
Equation of State (1979)

SI
17 - 31

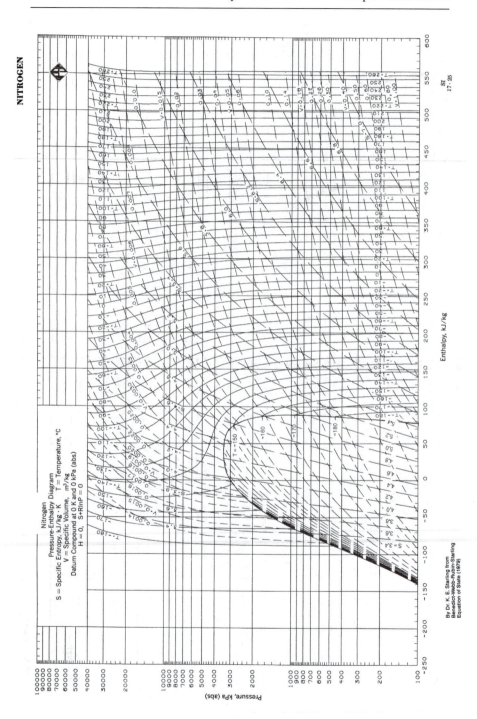

NITROGEN

Nitrogen
Pressure-Enthalpy Diagram
S = Specific Entropy, kJ/kg · K   T = Temperature, °C
V = Specific Volume, m³/kg
Datum Compound at 0 K and 0 kPa (abs)
H = 0, S+RlnP = 0

Pressure, kPa (abs)

Enthalpy, kJ/kg

SI
17 - 25

By Dr K. E. Starling from
Benedict-Webb-Rubin-Starling
Equation of State (1979)

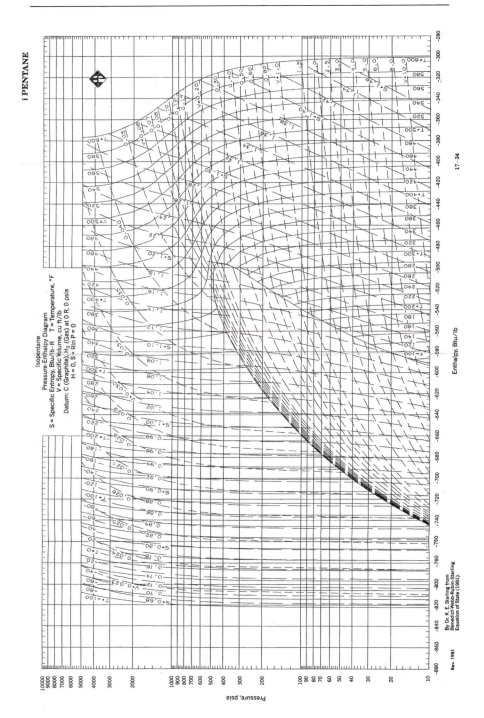

i PENTANE

Isopentane
Pressure-Enthalpy Diagram
S = Specific Entropy, Btu/lb·R   T = Temperature, °F
V = Specific Volume, cu ft/lb
Datum: C (Graphite), H₂ (Gas) at 0 R, 0 psia
H = 0, S + R in P = 0

Enthalpy, Btu/lb

Pressure, psia

By Dr. K. E. Starling from
Benedict-Webb-Rubin-Starling
Equation of State (1981).

Rev. 1981

17 · 34

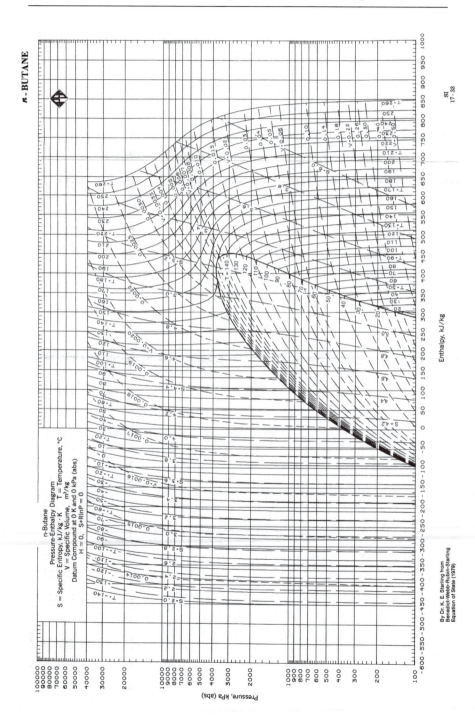

n - BUTANE

n-Butane
Pressure-Enthalpy Diagram
S = Specific Entropy, kJ/kg · K   T = Temperature, °C
V = Specific Volume,   m³/kg
Datum Compound at 0 K and 0 kPa (abs)
H = 0,   S+RlnP = 0

Enthalpy, kJ/kg

Pressure, kPa (abs)

By Dr. K. E. Starling from
Benedict-Webb-Rubin-Starling
Equation of State (1979)

SI
17 - 33

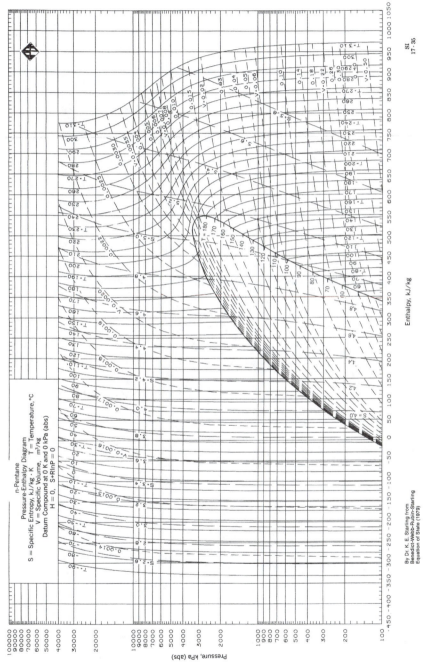

n-PENTANE

n-Pentane
Pressure-Enthalpy Diagram
S = Specific Entropy, kJ/kg · K   T = Temperature, °C
V = Specific Volume, m³/kg
Datum Compound at 0 K and 0 kPa (abs)
H = 0, S+RlnP = 0

Enthalpy, kJ/kg

Pressure, kPa (abs)

By Dr. K. E. Starling from
Benedict-Webb-Rubin-Starling
Equation of State (1979)

SI
17 - 35

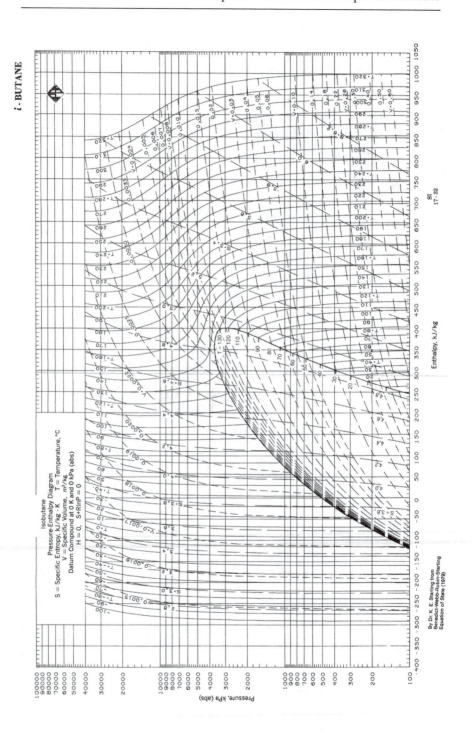

i - BUTANE

Isobutane
Pressure-Enthalpy Diagram
S = Specific Entropy, kJ/kg · K   T = Temperature, °C
V = Specific Volume, m³/kg
Datum Compound at 0 K and 0 kPa (abs)
H = 0, S+RlnP = 0

Enthalpy, kJ/kg

Pressure, kPa (abs)

SI
17 - 32

By Dr. K. E. Starling from
Benedict-Webb-Rubin-Starling
Equation of State (1979)

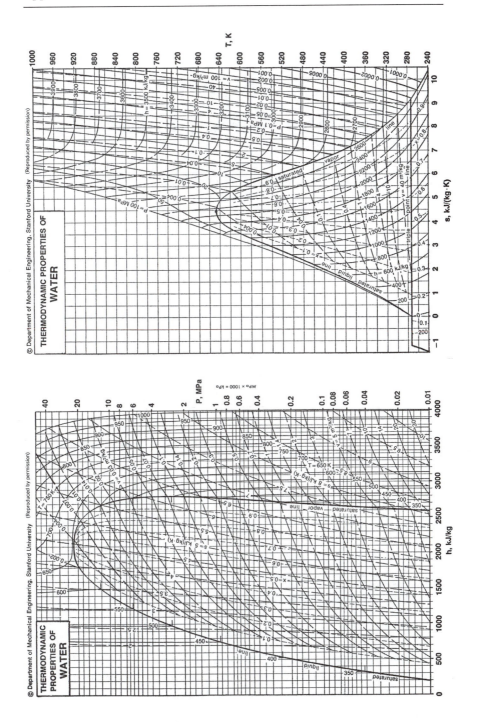

Appendix D

Classification of Hazardous Atmospheres

Area classifications are divided into CLASS, DIVISION, and GROUPS. As shown in the following table:

CLASS		DIVISION		GROUP	
1	Gases, Vapors	1	Normally Hazardous	A	Acetylene (581 ^{0}F)
				B	Hydrogen (968 ^{0}F)
				C	Ethylene (842 ^{0}F)
				D	Fuels such as: Methane (999 ^{0}F) Propane (842 ^{0}F) Butane (550 ^{0}F) Gasoline (536-880 ^{0}F) Naphtha (550 ^{0}F)
1	Gases, Vapors		Not Normally Hazardous	A	Same as Div. 1
				B	Same as Div. 1
				C	Same as Div. 1
				D	Same as Div. 1
II	Combustible Dust	1	Normally Hazardous	E	Combustible metal dust
				F	Carbonaceous dust Resistivity > 10^2 and <10^8 ohm-centimeter
				G	Carbonaceous dust Resistivity > 10^8 ohm-centimeter
		2	Not Normally Hazardous	F	Carbonaceous dust Resistivity > 105 ohm-centimeter
				G	Same as Div. 1
III	Easily Ignitable Fibers and Flyings				

The above information has been extracted from NFPA 70-1990, National Electric Code. National Electric Code and NEC are Registered Trademarks of the National Fire Protection Association, Inc. Quincy, MA.

Appendix E

Air/Oil Cooler
Specifications Check List

1. Site Conditions
    A. Location _____
    B. Altitude _____

C. Ambient Temperature Range
    Max _____ Min _____

2. Type
    A. Vertical _____   Horizontal _____
    B. Indoor _____   Outdoor _____

C. Louvers: With _____   Without _____
    Drives: Electric _____   Hydraulic _____

3. Number of Cooling Sections _____

4. Section _____ Gas Generator
    A. Oil Flow _____gpm
    B. Heat Loa d_____btu/hr
    C. Design Temp 300 F (150 C)
        Other _____
    D. Design Press._____psig

    E. Oil Outlet Temp 140 F (60 C)
        Other _____
    F. Allowable ΔP __5 psig__ (oil Side)
    G. Type Oils: _____
        Specific Gravity _____

5. Section _____ Power Turbine
    A. Oil Flow_____gpm
    B. Heat Load_____btu/hr
    C. Design Temp 300 F (150 C)
        Other _____
    D. Design Press._____psig

    E. Oil Outlet Temp 140 F (60 C)
        Other _____
    F. Allowable ΔP __5 psig__ (oil Side)
    G. Type Oils: _____
        Specific Gravity _____

6. Codes
    A. Design to ASME
        Stamp _____Yes _____No
    B. Section VIII Boiler Code for
        Location _____

    C. API Std __661__
    D. Customer Spec _____

7. Electrical Requirements
    A. Codes UL_____ CSA _____
    B. Hazardous Area Classification
        Class ___, Div ___ Group ___

    C. Available Power
        AC_____, Ph _____, Hz ____
        DC _____
        Other _____

8. Fan Requirements
    A. Non-Sparking ___ Yes ___ No

    B. Manually Adjustable Pitch Blades
        ____ Yes ____ No

9. Drive Requirements
    A. Hydraulic Drives
        _____ Direct Connected

    B. Electric Drive Connection
        ____ Direct _____ Indirect
    Indirect Electric Drive _____belt _____gear

10. Vibration Switch - Remote Resetable
    SPDT _____ (for hydraulic drives)

    DPDT _____ (for electric drives)

10. Vibration Switch - Remote Resetable
      SPDT _____ (for hydraulic drives)   DPDT _____ (for electric drives)

11. Special Conditions
      A. Physical Location   D. Actuator Temperature Control Switches _____
         Wall Mounted _____ Floor _____   E. Outside Louvers To Be Weather Tight _____
      B. Louvers Required   F. Air Actuators To Be Tubed To Coupling On
         Inside _____ Outside _____   Radiator _____
      C. Louver Actuators - When Required To
         Be Fail-safe, Closing Outside Louvers

12. Paint
      A. Per Customer Spec _____   C. Vendor Standard _____
      B. Per Manufacturer Spec _____   D. Galvanized _____ Strainless Steel _____

13. Electrical Conditions
      A. Actuators, temperature controls, vibration switches to be wired to external junction
         boxes to met electrical codes.
      B. Electric motor drives (and motor heater) to include conduit wiring to external surface of
         radiator enclosure.

14. Noise
      A. Noise levels to meet customer spec for   B. Outside source noise to be provided by
         field requirement _____   Purchaser _____

15.   Hinged Access Door To Inlet Side Of Radiator Is Required Unless Otherwise Specified

16.
Provision for floor drain of radiator enclosure box _____ Yes _____ No

17. Support Legs _____ Yes _____ No

18. Tube Bundles To Be
      Cleanable _____   Material: Steel with Aluminum Fins,
      Turbulators Removable _____   Other_____
      Type Of Fin Construction: Fouling Factor 0.0005
         Rolled In   _____   Std/Customers Spec _____
         Tension Wound _____
         Welded "L"   _____
         Other   · _____

19. Hydraulic Pressure Available _____ psig
                  Other _____ psig

20. Glycol/Water Heating Coil By Spec _____ (if required)

21. Insulation By Spec _____ (if required)

22. Packaging For Shipment Per Spec _____

23. Spare Parts List Required _____ Yes _____ No

24. Commercial Considerations
      A. Delivery Committment _____   C. Warrantees Std _____ Other _____
      B. Progress Reports __ Yes __ No   D. Sourcing _____
         Monthly _____ Other _____

25 Drawings
    A. Proposal Outline _____
    B. Schedule For Approval Drawings (from date of order) _____
    C. Schedule For Final Certified Drawings (from date of order) _____

26 Quality and Inspection
    A. Witness and Inspection Notice _____
    B. Shipping Release _____
    C. Material Certification _____
    D. Radiographic Requirements _____

27. Installation and Maintenance Manuals
    A. Number Required _____
    B. Delivery Required _____

Appendix F

Cylinder Displacement Curves

The following charts are provided as a shortcut method of determining cylinder displacement as a function of piston stroke and cylinder diameter. Curves are provided in both SI units and English units.

The charts have been provided courtesy of Petroleum Learning Programs, 305 Wells Fargo Drive, Suite 4, Houston, Texas 77090. Phone: (281) 444-7632, Fax: (281) 586-9876. E-mail: PetroLearning@aol.com.

SMALL DIAMETER COMPRESSOR CYLINDER DISPLACEMENT - SI UNITS

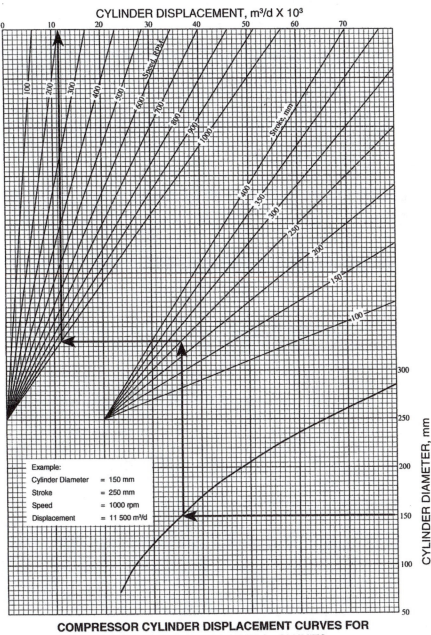

**COMPRESSOR CYLINDER DISPLACEMENT CURVES FOR
SMALL DIAMETER CYLINDERS - SI UNITS**

LARGE DIAMETER COMPRESSOR CYLINDER DISPLACEMENT - SI UNITS

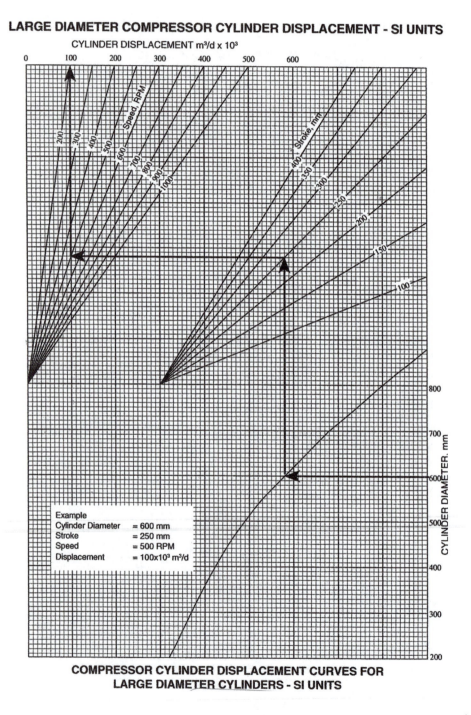

CYLINDER DISPLACEMENT m³/d x 10³

Example
Cylinder Diameter = 600 mm
Stroke   = 250 mm
Speed   = 500 RPM
Displacement   = 100x10³ m³/d

CYLINDER DIAMETER. mm

COMPRESSOR CYLINDER DISPLACEMENT CURVES FOR
LARGE DIAMETER CYLINDERS - SI UNITS

SMALL DIA. COMPRESSOR CYLINDER DISPLACEMENT — ENGLISH UNITS

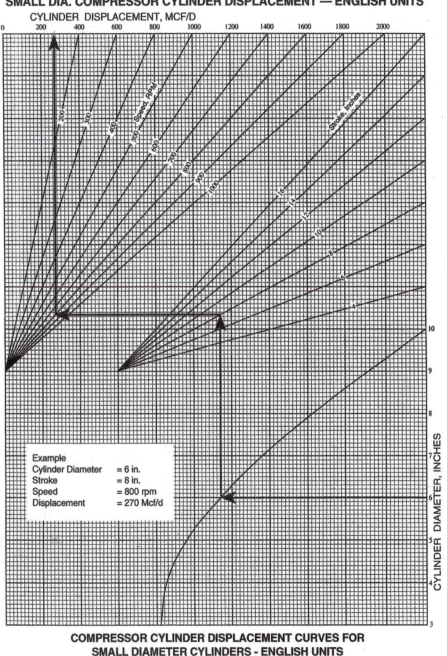

**COMPRESSOR CYLINDER DISPLACEMENT CURVES FOR
SMALL DIAMETER CYLINDERS - ENGLISH UNITS**

LARGE DIA. COMPRESSOR CYLINDER DISPLACEMENT - ENGLISH UNITS

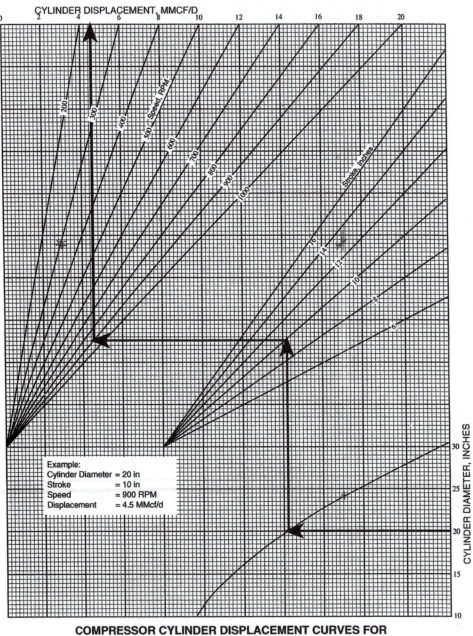

**COMPRESSOR CYLINDER DISPLACEMENT CURVES FOR
LARGE DIAMETER CYLINDERS - ENGLISH UNITS**

Appendix G

Compressor Cylinder Lubrication

The following charts are provided to determine cylinder lubricating oil flow as a function of piston stroke, cylinder diameter, speed and the number of oil injection ports. Curves are provided in both SI units and English units.

The charts have been provided courtesy of Petroleum Learning Programs, 305 Wells Fargo Drive, Suite 4, Houston, Texas 77090. Phone: (281) 444-7632, Fax: (281) 586-9876. E-mail: PetroLearning@aol.com.

LUBRICATION SYSTEM

COMPRESSOR CYLINDER LUBRICATION - ENGLISH UNITS

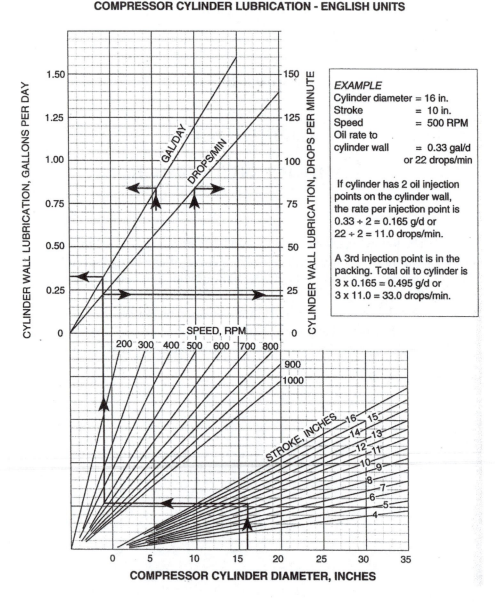

EXAMPLE
Cylinder diameter = 16 in.
Stroke   = 10 in.
Speed   = 500 RPM
Oil rate to
cylinder wall   = 0.33 gal/d
                  or 22 drops/min

If cylinder has 2 oil injection
points on the cylinder wall,
the rate per injection point is
0.33 ÷ 2 = 0.165 g/d or
22 ÷ 2 = 11.0 drops/min.

A 3rd injection point is in the
packing. Total oil to cylinder is
3 x 0.165 = 0.495 g/d or
3 x 11.0 = 33.0 drops/min.

LUBRICATION SYSTEM

COMPRESSOR CYLINDER LUBRICATION - SI UNITS

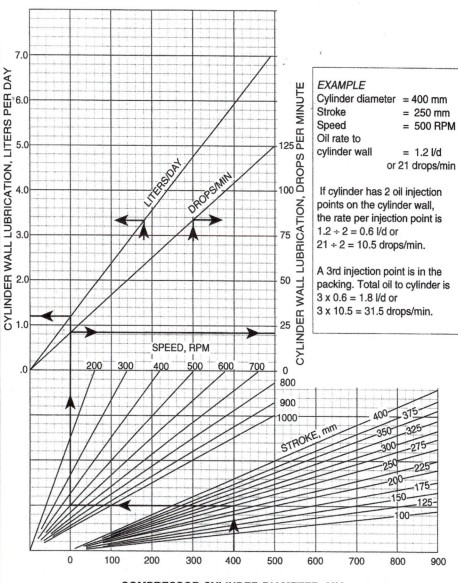

EXAMPLE
Cylinder diameter = 400 mm
Stroke   = 250 mm
Speed   = 500 RPM
Oil rate to
cylinder wall   = 1.2 l/d
                               or 21 drops/min

If cylinder has 2 oil injection
points on the cylinder wall,
the rate per injection point is
$1.2 \div 2 = 0.6$ l/d or
$21 \div 2 = 10.5$ drops/min.

A 3rd injection point is in the
packing. Total oil to cylinder is
$3 \times 0.6 = 1.8$ l/d or
$3 \times 10.5 = 31.5$ drops/min.

COMPRESSOR CYLINDER DIAMETER, MM

Appendix H

Troubleshooting Chart
Positive Displacement Compressors

The following is a list of symptoms or problems that most frequently occur. The possible causes associated with each problem are listed with the most likely cause or causes first.

Sympton or Problem	Compressor Type	Possible Cause
Compressor Driver Fails to Start	All	a) Control malfunction b) Low or no oil pressure
Compressor Fails To Rotate	Recip Recip All	a) Cylinder pressurized b) Seized components (crankshaft, piston, crosshead, bearings) c) Control malfunction
No Crankcase Oil Pressure	All	a) Oil filter clogged b) Oil system not primed c) No oil in crankcase d) Air leak in Pump inlet line e) Oil pump assembly defective
Low Crankcase Oil Pressure	All	a) Worn bearings b) Oil pressure relief valve spring has inadequate spring tension c) Oil pressure relief valve plugged d) Low crankcase oil level e) Plugged suction strainer
High Crankcase Oil Pressure	All	a) Oil pressure relief valve spring has excessive spring tension b) Plugged oil pressure line
High Crankcase Oil Temperature	All	a) Cooling inadequate b) Low oil level c) Oil contaminated d) Improper oil
No Gas Throughput	All Recip All	a) Controls inoperative b) Missing or defective valves c) Restricted suction line
Reduced Gas Throughput	All Recip Recip All Recip	a) Restricted suction line b) Worn, loose or missing valves c) Piston rings worn d) Control not operating property e) Open or defective VVP or suction valve unloader
Cylinder Overheating	 Recip	a) Cooling insufficient b) Contamination in cooling water jacket or valve assemblies c) Broken valves or valve springs d) Piston rod packing too tight e) Piston, liner or cylinder wall scored f) Discharge pressure too high g) Suction pressure too low h) Insufficient lubrication
High Discharge Pressure	All All	a) Control setting improper b) Discharge line partially blocked
Low Discharge Pressure	All Recip Recip All	a) Control system defective b) Worn, loose or missing valves c) Worn piston rings d) Compressor capacity too low relative to system demand

SYMPTOM	Compressor Type	POSSIBLE CAUSE	
High Discharge Temperature	Recip	a)	Worn, loose or missing valves
	All	b)	Cooling insufficient
	All	c)	Suction temperature too high
Broken Valves & Valve Springs	Recip	a)	Incorrect assembly
	Recip	b)	Foreign material or liquids
	Recip	c)	Insufficient lubrication
Scored Cylinder Liner & Piston	Recip	a)	Foreign material in cylinder
	Recip	b)	Broken valves & springs
	Recip	c)	Cylinder out-of-round
	Recip	d)	Cylinder temperature too high
	Recip	e)	Inadequate lubrication
High Intercooler Pressure (where applicable)	Recip	a)	Broken or missing 2nd stage (or higher) inlet valves, discharge valves
	Recip	b)	Valve cage misaligned
	Recip	c)	Worn piston rings in successive stages
	Recip	d)	Worn cylinder liner in successive stages
	All	e)	Suction pressure too high
Low Intercooler Pressure	Recip	a)	Broken or missing 1st stage inlet valves, discharge valves
		b)	Piping leak in intercooler or interconnecting pipe
	All	c)	Intercooler drain malfunction
	All	d)	Worn 1st stage piston rings
	All	e)	Piston rod packing leak
	Recip	f)	Suction pressure too low
	All		
Knocking	Recip	a)	Loose & cage
	Recip	b)	Worn crosshead or crosshead liner
	Recip	c)	Worn connecting rod bearings
	Recip	d)	Liquids or foreign material in cylinder
	Recip	e)	Head clearance incorrect
	Recip	f)	Piston loose on piston rod
	Recip	g)	VVP plug loose
Compressor Vibration	All	a)	Compressor out of balance
	All	b)	Installation or alignment improper
	All	c)	Worn bearings
	Recip	d)	Loose valve assemblies
Compressor Performance Decrease	All	a)	Compressor fouling
	All	b)	Foreign object damage
	All	c)	Worm seals
Compressor Vibration Imbalance	All		Foreign object damage
	All		Compressor fouling
Compressor Axial Vibration	All	a)	Coupling Misalignment

Appendix I

Starting, Operating and Maintenance Procedures

CONTENTS
- Preparation for Start-up
- Operation
- Maintenance

PREPARATION FOR START-UP

Compressor

On units shipped with the cylinders and piston assemblies installed, it is necessary to check piston end clearances, piston rod run out, crosshead shoe clearances, and piston pin through-bolt torque and correct if necessary. Record these readings on the frame setting record sheet for future reference.

Remove the suction valves from each end of the compressor cylinders to permit the unit to run at no load for break in purposes. Reinstall and tighten the valve covers.

Remove the compressor crankcase cover and cross head inspection doors on the compressor. Inspect the compressor crankcase to insure that it is free of all dirt, water and long-term preservatives. Inspect the compressor oil filter cases for dirt, rust, etc. Clean the filter cases if needed.

Fill all cooling systems with coolant and bleed all air from the system. Allow the unit to sit for 2-3 hours then check inside the compressor cylinders for coolant leaks.

If not already installed, install the compressor oil filter element and filter top using a gasket if needed. Consult the instruction manual for the correct type and weight of oil to be used in the compressor crankcase.

Install oil in the compressor crankcase to the proper level. Using the prelube hand pump, prelube the compressor frame. Bleed all air from the compressor oil filter. Continue to prelube the compressor until oil is delivered to all lubrication points of the compressor frame. A visual check will determine if oil is reaching the main bearing, rod bearings, and crosshead assemblies.

After it has been established that the frame is lubricated properly, reinstall the crankcase top covers and the crosshead inspection doors.

Inspect the compressor lubricator reservoir for water that may have entered during shipping and storage. Clean the reservoir if needed. Check the lubricator drive for proper alignment and ease of operation. This is done by moving the lubricator oil pump coupling back on the oil pump shaft and rotating the oil pump by hand. If binding is found to exist, loosen the lubricator mounting bolts and move the lubricator from side to side or up and down until unit turns freely. Tighten lubricator mounting bolts and connect the compressor oil pump drive coupling to the oil pump drive shaft.

If the lubricator is not fed from the compressor frame be sure it is connected to a supply with the proper type and weight of oil, as recommended in the instruction manual, for the break-in of the compressor cylinders and packings and for the type of operation under which the compressor will be run.

Using the hand primers on lubricator pump or a suitable hand primer gun, prime and bleed all lubricator lines until they are free of air. Continue to hand prime the system until there is oil in the cylinder bore and packing cases.

Motor Check

See the motor section of the instruction manual for the manufacturer's instructions for start-up.

Auxiliaries

Fill all systems that require oil such as the mist lubricator for the starter and air/gas prelube pump motors.

Grease the cooler fan shaft and idler bearings, if needed.

Control Panel

Make a preliminary inspection of the operation of the control panel. During the break in period of the unit it may be necessary to "lock out" certain parts of the control system in order to keep the unit running. This depends on the type of panel used on the unit. Do not, under any circumstances, lock out the low oil pressure shutdowns or completely disable the panel and run the unit. Should a malfunction occur on the compressor during the break-in run, serious damage could occur before the unit could be shutdown by the plant operator.

See the panel section of the instruction manual for complete details.

Instrumentation

A point-to-point continuity check must be made for all wiring. Conduit seal fittings (if applicable) must be filled prior to start-up. Any instrumentation which was removed for shipping should be re-installed at this time.

Start-up Procedure

Start the unit and run for 5 minutes. Do not exceed the maximum number of motor starts per hour. Remove the compressor crankcase covers and check for any overheating of the crankcase main bearings rod bearing and crosshead assemblies. Reinstall the crankcase covers if no heating is found.

Restart the unit and run for 30 minutes. Note the compressor oil pressures, oil temperatures and monitoring device for main bearing temperatures if provided. Stop the unit and repeat the check for overheating of the compressor bearings, (and motor bearings) and other items listed in step 1.

Remove the suction valve covers and inspect the cylinder bore for oil, scuffing, and debris. Check the packing glands for overheating.

If there is no indication of overheating of the crankshaft, cylinder bores or packing glands, restart the unit and run for a second thirty minute period. Note the oil pressures and oil & water temperatures.

At the end of the second thirty minute period, stop the unit. Make a final inspection of the compressor crankshaft, main bearings, rod bearings, motor bearings, crosshead assemblies, rod packings and cylinder bores for overheating and proper lubrication. Inspect the bores of all cylinders for a glazing effect. If everything is found to be satisfactory, the unit can now be loaded. Install the suction valves and valve covers, and the crankcase cover.

After all valve covers, piping, etc., have been installed, open the unit suction valve and allow pressure to flow through the unit out the blow down valve until the unit is adequately purged. Once all air is expelled, allow the pressure to build to 50 psi or normal suction· pressure whichever is less. It is suggested that air be purged from the unit using nitrogen gas. Once all the air is adequately purged, then the nitrogen may be purged from the unit using the process gas.

Lubrication & Cooling

The complete lubricating system of the compressor may be conveniently divided into three parts of equal importance. Complete protection of all frame running parts is provided by lube oil from the frame sump. A separate, independent, force feed lubricator and tubing system provides lubrication for cylinder walls and piston rod packing. A shell-and-tube cooler is provided for frame oil cooling.

Lubricating Oil Requirements

A good mineral oil which provides resistance to oxidation and corrosion is generally satisfactory for lubrication in a reciprocating compressor which has its crankcase sealed off from the cylinders. However, there is no objection to the use of a detergent type oil if this is more readily available. The best assurance of obtaining a suitable oil is to use only products of well known merit, produced by responsible concerns, and used in accordance with their recommendations. Do not permit your compressor to be used as an experimental unit for trying out new or questionable lubricants.

In some cases it may be convenient or practical to use the same type oil in the compressor as is used in the compressor drive engine. This is permissible as long as the engine oil is of proper viscosity.

If start-ups are to take place when ambient temperatures are below freezing, the **pour point** of the oil must be low enough to insure flow to the oil pump. Heavier oil should be heated before starting.

If a compounded oil is used, the non-corrosiveness of this oil must be looked into very carefully. The oil must not contain substances which might be injurious to tin or lead base babbitts and it is also highly desirable that it be non-corrosive to copper-lead alloys.

Mechanical Check List

1. Compressor jack screws are backed off after grout has set up
2. Compressor anchor bolts are properly torqued.
3. Check Coupling Alignment.
4. Motor anchor bolts are properly torqued.

5. Check tightness of adapter plate bolts to flywheel and sheave (if applicable).
6. After cylinder is mounted on compressor. Jam nuts are properly tightened.
7. Check packing gland for clearance. (Record Clearance)
8. Packing bolts are tight.
9. Torque crosshead pin bolts. Procedure to torque these bolts are:
   a. Torque the bolts
   b. Tap the bolt with a hammer
   c. Re-torque the bolt.
10. Check and record cross head clearance.
11. Record piston end clearance.

OPERATION

Preparation for Initial Start-up

CAUTION

READ THIS PROCEDURE AND FAMILIARIZE YOURSELF WITH THE COMPRESSOR AND ITS AUXILIARY EQUIPMENT PLUS YOUR COMPANY SAFETY PROCEDURES BEFORE ATTEMPTING TO START THIS MACHINERY.

The following procedure is suggested before starting the unit· for the first time, or after overhaul, or after a long period of idleness.

1. Remove the top cover of the base and the covers for the crossheads and distance pieces on each crosshead guide. Check for cleanliness before adding lube oil to the base. Be sure that no dirt, cuttings, water, etc., are allowed to remain.
2. Remove a valve from each end of each compressor cylinder. Rotate the compressor slowly and check for piston-to-head clearances. Make sure all parts move freely. Check that all valves are installed property.

CAUTION

BEFORE REMOVING ANY GAS CONTAINING PART OF THE COMPRESSOR OR ASSOCIATED PIPING SYSTEM, VENT THE COMPRESSOR AND SYSTEM TO ATMOSPHERE.

3. Add lubricating oil to the base and to the lube oil filter.

4. Check the force feed lubricator for cleanliness and fill to the proper level with oil.

5. Adjust all force feed lubricator pumps to full stroke for cylinder and packing break-in.

6. Disconnect ends of force feed lubricator lines as close as possible to cylinders and cross head guides. Hand pump the lubricators to fill lines and eliminate air.

7. Connect the force feed lubricator lines and operate pumps ten more strokes to force oil into cylinders and rod packing.

CAUTION

HIGH PRESSURE OIL STREAM MAY PUNCTURE SKIN. USE PROPER WRENCH AND KEEP HANDS AWAY FROM IMMEDIATE POINT WHERE CONNECTION IS PURGING AIR.

8. Prime the system with the lube oil priming pump.

9. Hand lubricate the piston rod next to the packing.

10. Replace all covers with their respective gaskets and tighten screws according to torque chart. Distance piece covers may be left off to check for packing leaks on start up if gas is not sour gas.

11. Check to see that all crosshead guides or distance pieces and packings are vented individually and with proper size vent lines.

12. Unload the compressor for start-up.

13. Check to be sure all guards are in place.

Initial Start-up

1. Open the valves supplying water to the compressor cooling system (when required).

2. Start up and operate the unit under no-load conditions at reduced speed where possible (i.e., engine driven units). Check the oil pressure. When the compressor is started, an oil pressure of 20 psi must be experienced within 5 seconds or the compressor must be immediately shut down. Do not restart until adequate oil pressure can be assured.

3. After running the unit 15 to 20 minutes, shut down and check all bearings and packings for high temperature.

4. Check piping for oil or water leaks.

5. Start up again and run for approximately 20 to 30 minutes. Add oil to the crankcase to bring the oil level (while running) up to the

middle of the sight glass. Shut down and recheck as above.

6.   Start the unit and bring it up to full rated speed. Apply the load.
7.   During the initial period of operation, pay close attention to the machine for any unusual high temperature, pressure, or vibration. In the event of equipment malfunction where excessive vibration, noise, high temperature, or any other dangerous condition exists, the compressor should be stopped immediately.

WARNING

IF COMPRESSOR HAS BEEN STOPPED, DO NOT IMMEDIATELY RE-MOVE EQUIPMENT COVERS. ALLOW THE UNIT TO COOL DOWN TO PREVENT POSSIBLE EXPLOSION DUE TO IN RUSH OF AIR, AND TO PREVENT INJURY WHICH MAY BE CAUSED BY HOT SURFACE.

Normal Start-up

Not all of the instructions provided for initial start-ups are required for routine starting. The following notes comprise the normal starting procedure:

1.   Unload the compressor.
2.   Operate the force feed lubricator pumps, by hand, for ten strokes. (Be sure the lubricator tank is kept full).
3.   Hand prime the frame lube oil system.
4.   Turn on cooling water supply.
5.   Start the unit. Check frame lube oil pressure.
6.   Operate at low speed (where possible) and no load for several minutes. Check force feed lubricator sight glasses for feed. Check lube oil for proper level at sight gauge.
7.   Bring up to rated speed and apply load.

Normal Shutdown

Before shutting the unit down, it is good practice to unload the compressor and reduce speed (where possible).

Emergency Shutdown

WARNING

IF COMPRESSOR HAS BEEN STOPPED, DO NOT IMMEDIATELY RE-MOVE EQUIPMENT COVERS. ALLOW THE UNIT TO COOL DOWN TO PREVENT POSSIBLE EXPLOSION DUE TO IN RUSH OF AIR, AND TO PREVENT INJURY WHICH MAY BE CAUSED BY HOT SURFACE.

Recommended Operating Conditions

For compressor operating speed ranges, refer to Compressor Performance Data Sheet.

WARNING

IMPROPER SETTING OF VARIABLE VOLUME POCKETS, FIXED VOLUME POCKETS, VALVE UNLOADERS OR OTHER UNLOADING DEVICES CAN RESULT IN DANGEROUS OPERATION AND POSSIBLE DAMAGE AND/OR INJURY TO EQUIPMENT OR PERSONNEL OPERATING THIS UNIT WITHOUT CLEARANCE AND LOADING INFORMATION CAN RESULT IN FAILURE DUE TO. OVERLOADING, EXCESSIVE ROD LOADS OR HIGH TEMPERATURE.

MAINTENANCE

General

The diligent observation of the inspection and maintenance procedure, given in this section, will go a long way toward insuring satisfactory operation of the compressor. Good preventative maintenance practice includes a periodic check of critical bolt torques, such as compressor main and connecting rod bolts and flywheel & sheave bolts. The recommended schedule should be followed:

1. One (1) month after the unit is placed into service.
2. Six (6) months after the unit is placed into service.
3. At least once every twelve (12) months thereafter.

This schedule should be repeated after each compressor overhaul. During the first 300 hours of operation, the compressor should be checked frequently. Check bearings and packings for excessively high temperatures, and look over the complete unit for oil or water leaks.

Constant care in maintaining the machine in clean condition will payoff in time, labor, and repair costs. Inspect and clean the unit at regular intervals.

When repair or inspection inside the covers of the frame or cylinders is required, use only clean, lint-free rags. Lint or loose threads may cause clogging of lubrication lines or filters and result in major troubles. All tools and work places should be kept free from grit, dust or dirt.

WARNING

WHEN WORK IS BEING DONE ON THE COMPRESSOR, THE ELECTRIC MOTOR MUST BE BLOCKED IN SUCH A WAY THAT THE COMPRESSOR CANNOT TURN OVER. VALVES MUST BE CLOSED ON THE SUCTION AND DISCHARGE LINES. AIR OR GAS MUST BE BLED OFF FROM THE CYLINDERS. PRECAUTION MUST BE TAKEN TO PREVENT THE OPENING OF ANY VALVE WHICH WOULD RELEASE PRESSURE AGAINST A PISTON, CAUSING IT TO ROTATE THE UNIT AT A CRITICAL MOMENT.

The following paragraphs contain information about the various assemblies that go to make up the complete compressor assembly. The conscientious operator will obtain valuable knowledge by a thorough reading of this material from time to time.

BASE (Crankcase)

The base is made of high strength iron and is heavily reinforced for maximum rigidity. Removable top and end covers provide access to all moving parts and the open top design allows easy crankshaft removal. Drilled lubrication passages in the base should be carefully cleaned at overhaul.

Crankshaft and Main Bearings

The complete crankshaft assembly includes the oil slinger. The crankshaft is rifle-drilled to carry lubrication from the main bearing to the connecting rod bearing. Plain and thrust main bearing shells are of the split, non-adjustable, precision type.

New upper and lower plain main bearing shells are interchangeable. However, after the compressor has been run, it is preferable that the shells be placed back in their original position after removal for any cause. Therefore, the bearing shells should be so marked.

NOTE
Use pencil markings only on the parting line faces of
bearing shells or in the bearing grooves.

Carefully clean the crankshaft and bearing shells and saddles before attempting to replace the bearing shells. Under no circumstances should any filing, scraping, or other fitting be done on either bearing shells or saddles. The bearing cap nuts should be tightened uniformly (criss-cross method) to the proper torque.

Connecting Rods

The connecting rod is a steel forging, rifle-drilled to provide lubrication to the crosshead pin bushing. The cap and rod are match-marked, and a complete assembly must be ordered if replacement is necessary.

The upper end of the connecting rod carries a pressed-in bushing. This bushing has a large oil groove both inside and outside at its midpoint, with communicating holes to admit oil from the connecting rod to the crosshead pin. The inside diameter of the bushing has equally spaced, helical oil spreader grooves to insure adequate lubrication between bushing and crosshead pin. This bushing is precision bored after assembly in the rod to insure proper size and location. If these bushings are replaced in the field, extreme care should be used in maintaining the bore of the new bushings parallel to and properly spaced from the crankpin bore.

THE CAPS AND RODS ARE MATCH-MARKED. NUMBERED BY THROW AND HAVE THEIR RESPECTIVE WEIGHTS STAMPED ON. ALWAYS INSTALL RODS WITH THIS INFORMATION UP. The rod cap is removed and replaced while the crank throw is in the outer position and the rod itself can be eased into or out of position with the crank throw slightly below the inner position.

Crosshead Guides

The crosshead guides have lubrication holes with metering fittings for both top and bottom slides. BLOCKING OF THESE FITTINGS MAY CAUSE THE CROSSHEAD SHOES TO HEAT UP AND SEIZE.

The nuts and capscrews holding the guide to the base must be torqued evenly (criss-cross method) to prevent cocking of the guide relative to the base and crankshaft. Large side covers on the guide allow easy access to the crosshead, connecting rod, and rod packing. The crosshead can be removed through these openings without disturbing the cylinder mounting.

Crosshead—Removal and Installation

The crosshead is made of ductile iron and has removable top and bottom steel backed shoes with durable bearing material on the sliding surface. Flat head screws and locknuts hold the shoes firmly in place, and these must be torqued evenly. Like the main bearings maintenance, cleanliness is an important factor during the assembly of shoes to crosshead and crosshead to guide.

Cylinder Body

The cylinder body is provided with drilled water passages, top and bottom, which connect the water inlet and outlet with the cooling source. Whenever the water jacket covers on the cylinder sides are removed to clean out deposits, the drilled passages should also be cleaned out. If the pipe plugs in the crank end of the drilled passages are removed, they should be coated with good waterproof sealer and replaced. This will prevent water seepage into the atmospheric vent space.

Lube oil, from the force feed lubricator tubing system, passes through a check valve and into a fitting on the outside of the upper flange of the cylinder. From here, the flow is through the wall of the discharge flange, into a tube and fitting assembly which connects to a drilled passage near or at the top of the cylinder bore. Cylinders have two point lubrication, one in the upper flange and one in the lower flange. During general overhaul the oil hole for cylinder bore lubrication should be cleaned out and all steel tubing checked for soundness and tightness.

Cylinder Head

The gasket, which is used to seal the cylinder head to the cylinder body, may be re-used only if a new one is not available. However, if possible, a new gasket should be fitted if the head has been removed after a period of running. The water seal grommets should also be replaced at the same time.

Pistons, Piston Rings and Piston Rod

The piston to rod nut is torqued after locking two crosshead nuts on the crosshead end of the rod and clamping them in a large vise. (If the piston is unusually heavy, provide some kind of support during this operation.). DO NOT clamp onto the piston rod. The piston nut may also be torqued when rod and piston are in place and the rod is locked in the crosshead.

The piston end clearance is adjusted by mounting the head and measuring the clearance on both ends with a feeler gauge. Check these clearances with the crosshead nut snugged down. If adjustment is necessary, loosen crosshead nut, remove cylinder head and turn piston and rod assembly. Once again DO NOT use wrenches on rod shank; do all turning of piston rod through the piston to rod nut.

Piston rings are designed with proper end clearance. However, it is a good idea to check this in the cylinder. Rings should be square in the bore when checking.

Piston Rod Packing

The purpose of piston rod packing is to prevent the loss of gases from the cylinder along the piston rod. During initial operation a packing may leak or tend to overheat. This is a temporary condition while the rings are properly mating to the rod and case. As a general guide, a temperature in excess of that which can be tolerated by resting the hand on the packing flange will indicate too fast a rate of wearing in. The nominal rate of lubrication for the piston rod packing is specified by the manufacturer. However, definite lubrication rates and time intervals for wearing-in of the packing are difficult to prescribe here. Experience has indicated that these factors may vary widely on different applications. If there is concern for proper lubrication rate contact the manufacturer's representative.

Rod packing must be removed and replaced as a complete unit. This is possible after first removing the cylinder head, unlocking the crosshead nut and screwing the piston rod out of the crosshead. Leave the crosshead in the guide and bar compressor over until crosshead moves to inner dead center position to allow removal of the packing. Pull rod (with split sleeve over threads) through the packing until it clears. Disconnect tube fittings and remove screws holding packing to cylinder.

When disassembling a packing, always carefully note and record the position of each packing cup and each ring, and the direction each ring faces. Ordinarily, packing cups and glands are not subject to severe wear. It is, therefore, possible to make repairs by ordering only a set of new rings. It is always good practice to keep complete spare sets of new rings for the packing assembly.

Before installing a new packing assembly, it is important that the piston rod be carefully checked. If the rod is worn, rough, pitted, or has a taper, it must be replaced. The packing cups and gland, and all parts that are not replaced by new parts, must be soaked and thoroughly cleaned in a non-acid solvent. They should then be blown dry and examined closely for unusual nicks or burrs which might interfere with the rings free floating or contact with the rod. Particular care must be taken with rings made of soft metals and it is very important that wiper rings be handled and installed carefully to prevent damage to the scraping edges.

The rings must be placed in the packing cups in the same position (facing original direction) as the original set.

The stuffing-box bore for the rod packing must be cleaned and examined for burrs. If found, burrs should be carefully and completely removed. A new metallic gasket should be placed in the groove of the packing cup which seats in the stuffing box.

When installing a new or rebuilt packing assembly, the attaching screws must be tightened uniformly. The gasket must not cock or bind and the assembly must seat squarely in the bore.

NOTE

The packing case flange must not contact the cylinder body.
Clearance must exist at this point to ensure pressure is applied to
the seal ring when packing screws are tightened.

Before connecting the tubing to the packing flange, hand pump the force feed lubricator until oil runs from one of the disconnected tubes. Connect this tube to the respective hole in the packing flange and continue to pump lubricator 12 to 15 more strokes. Connect vent tubing. With split sleeve still on the rod threads, insert the piston rod (with piston still assembled thereon) back through the cylinder head opening and through the packing assembly. Remove the split sleeve, add the crosshead nut, and thread the piston rod back onto the crosshead.

Valve Installation

Suction and discharge valves must be installed in the proper direction. This can be determined by first inspecting the valve to see which direction the springs push the valve plates closed.

The gas will flow in the same direction that the valve plate moves when it compresses the spring toward the center of the cylinder. The discharge valves are installed with the springs away from the center of the cylinder.

Valve Replacement

WARNING

BEFORE REMOVING ANY GAS CONTAINING PART OF THE COMPRESSOR OR ASSOCIATED GAS PIPING SYSTEM, VENT COMPRESSOR AND SYSTEM TO ATMOSPHERIC PRESSURE.

Proceed with valve removal in the following manner:

1. Loosen bolts or nuts holding valve cap. DO NOT remove completely until cap is pulled out far enough to vent any pressure trapped under cap.

2. Remove valve cap. Inspect O-ring; replace if defective.

3. Loosen setscrew in valve retainer (bottom valves only); insert threaded puller into valve retainer and remove.

4. Using threaded valve puller, remove valve from seat in cylinder.

5. Remove gasket, inspect and replace as needed.

6. Clean the gasket surface on valve and in valve pocket. Use new or a good gasket.

7. Place valve in pocket, FACING PROPER DIRECTION.

8. Locate the retainer on top of the valve assembly and grease or oil the O-ring of the valve cap forcing the cap into place by torquing the nuts or screws evenly.

NOTE

The valve cap flange must not contact the cylinder body.
Clearance must exist at this point to insure that pressure is applied to the valve and retainer gaskets when the valve cap bolts or nuts are tightened.

To <u>replace a valve into a bottom port</u> (this is a discharge port), proceed as follows:

9. Invert retainer. Place valve on top of retainer with valve guard facing out (away from cylinder). Slip gasket on valve assembly.

10. Lift the complete works up into the bottom port, making sure that the valve seat enters first.

11. Tighten the retainer setscrew just enough to hold everything in place.

12. Lubricate valve cap O-ring.

NOTE: At the completion of any maintenance, it is important to bar the compressor crankshaft over, at least one complete revolution, to check for adequate running clearances.

RECOMMENDED COMPRESSOR MAINTENANCE SCHEDULE

Controls	Daily	Weekly	Monthly	Semi-Annually	Annually or As Needed
Perform safety shutdown checks				X	
Note and record panel gage readings	X				
Check calibration of all thermometers and pressure gages				X	

Lubrication System	Daily	Weekly	Monthly	Semi-Annually	Annually or As Needed
Check oil levels; oil level regulator, oil level sight gage	X				
Check lubricator line connections for leakage	X				
Inspect frame, lubricator and packing cases for leakage		X			
Clean and/or replace crankcase breather(s)				X	
Check compressor force feed lubricator pump(s) for proper quantity output				X	
Take compressor oil sample for analysis			X		
Change compressor oil and oil filters	Lubricating oil and filter elements should be changed after the initial 400 hours run time. Thereafter, change-out can be increased to 1000 hour intervals. Extend oil changes to longer intervals if the oil analysis company recommends continued use of oil. Change lube oil filters when a differential pressure of 12-15 psi has been reached.				
Replace O-rings					X

Mechanical/Operator System - Cylinder	Daily	Weekly	Monthly	Semi-Annually	Annually or As Needed
Note and record inlet temperatures	X				
Check for loose cylinder mounts, piping connections, supports		X			
Note and record cylinder discharge temperatures	X				
Hand check suction valve covers for coolness	X				
Listen for unusual noises	X				

Mechanical/Operator System – Cylinder (continued)	Daily	Weekly	Monthly	Semi-Annually	Annually or As Needed
Check temperatures of coolant to and from cylinders and lube oil cooler, packings	X				
Check one compressor valve of each stage, inspect for broken plates or springs, trapped liquids or solids				X	
Remove the distance cover piece or crosshead guide cover to inspect packing on each piston rod				X	
Remove the first stage head-end piston to check cylinder bore, piston rings, piston, rod and rod bearing condition					X

Mechanical/Operator System - Frame	Daily	Weekly	Monthly	Semi-Annually	Annually or As Needed
Check crosshead clearances					
Check crosshead guide for wear metals					
Check foundation bolt torques					
Check compressor coupling for proper alignment					
Visually inspect frame interior for bearing material in frame, gear tooth condition, crosshead shoe and guide condition					
Roll out compressor thrust lower main shell for inspection					
Check compressor accessory drive gear back-lash and condition					

Appendix J

Basic Motor Formulas
And Calculations

The following formulas are provided courtesy of Baldor Electric Company to assist the reader in selecting the proper motor driver for his compressor application.

The Baldor Electric Company is located at 5711 R.S. Boreham, Jr. St., Fort Smith, AR 72901. Phone: (479) 646-4711 ext.5123. Fax (479) 648-5281. Web Site www.baldor.com

BASIC MOTOR FORMULAS AND CALCULATIONS

The formulas and calculations which appear below should be used for estimating purposes only. It is the responsibility of the customer to specify the required motor Hp, Torque, and accelerating time for his application. The salesman may wish to check the customers specified values with the formulas in this section, however, if there is serious doubt concerning the customers application or if the customer requires guaranteed motor/application performance, the Product Department Customer Service group should be contacted.

Rules of Thumb (Approximation)

At 1800 rpm, a motor develops a 3 lb.ft. per hp

At 1200 rpm, a motor develops a 4.5 lb.ft. per hp

At 575 volts, a 3-phase motor draws 1 amp per hp

At 460 volts, a 3-phase motor draws 1.25 amp per hp

At 230 volts a 3-phase motor draws 2.5 amp per hp

At 230 volts, a single-phase motor draws 5 amp per hp

At 115 volts, a single-phase motor draws 10 amp per hp

Mechanical Formulas

$$\text{Torque in lb.ft.} = \frac{\text{HP x 5250}}{\text{Rpm}} \quad \text{HP} = \frac{\text{Torque x rpm}}{5250} \quad \text{rpm} = \frac{120 \text{ x Frequency}}{\text{No. of Poles}}$$

Temperature Conversion

$$\text{Deg C} = (\text{Deg F} - 32) \text{ x } 5/9$$

$$\text{Deg F} = (\text{Deg C x } 9/5) + 32$$

High Inertia Loads

$$t = \frac{WK^2 \text{ x rpm}}{308 \text{ x T av.}}$$

WK^2 = inertia in lb.ft.^{2}
t = accelerating time in sec.
T = Av. accelerating torque lb.ft..

$$T = \frac{WK^2 \text{ x rpm}}{308 \text{ x t}}$$

$$\text{inertia reflected to motor} = \text{Load Inertia} \left(\frac{\text{Load rpm}}{\text{Motor rpm}} \right)^2$$

Synchronous Speed, Frequency and Number of Poles of AC Motors

$$n_s = \frac{120 \text{ x f}}{P} \qquad f = \frac{P \text{ x } n_s}{120} \qquad P = \frac{120 \text{ x f}}{n_s}$$

Relation between Horsepower, Torque, and Speed

$$HP = \frac{T \text{ x } n}{5250} \qquad T = \frac{5250 \text{ HP}}{n} \qquad n = \frac{5250 \text{ HP}}{T}$$

Motor Slip

$$\% \text{ Slip} = \frac{n_s - n}{n_s} \text{ x } 100$$

Code	KVA/HP	Code	KVA/HP	Code	KVA/HP	Code	KVA/HP
A	0-3.14	F	5.0 -5.59	L	9.0-9.99	S	16.0-17.99
B	3.15-3.54	G	5.6 -6.29	M	10.0-11.19	T	18.0-19.99
C	3.55-3.99	H	6.3 -7.09	N	11.2-12.49	U	20.0-22.39
D	4.0 -4.49	I	7.1 -7.99	P	12.5-13.99	V	22.4 & Up
E	4.5 -4.99	K	8.0 -8.99	R	14.0-15.99		

Symbols

I   = current in amperes
E   = voltage in volts
kW = power in kilowatts
kVA = apparent power in kilo-volt-amperes
HP = output power in horsepower
n   = motor speed in revolutions per minute (RPM)
ns = synchronous speed in revolutions per minute (RPM)
P   = number of poles
f   = frequency in cycles per second (CPS)
T   = torque in pound-feet
EFF = efficiency as a decimal
PF = power factor as a decimal

Equivalent Inertia

In mechanical systems, all rotating parts do not usually operate at the same speed. Thus, we need to determine the "equivalent inertia" of each moving part at a particular speed of the prime mover.

The total equivalent WK^2 for a system is the sum of the WK^2 of each part, referenced to prime mover speed.

The equation says:

$$WK^2_{EQ} = WK^2_{part} \left(\frac{N_{part}}{N_{prime\ mover}} \right)^2$$

This equation becomes a common denominator on which other calculations can be based. For variable-speed devices, inertia should be calculated first at low speed.

Let's look at a simple system which has a prime mover (PM), a reducer and a load.

$$WK^2 = 900 \text{ lb.ft.}^2$$

$WK^2 = 100 \text{ lb.ft.}^2$ (as seen at output shaft) $WK^2 = 27{,}000 \text{ lb.ft.}^2$

PRIME MOVER   3:1 GEAR REDUCER   LOAD

The formula states that the system WK^2 equivalent is equal to the sum of WK^2_{parts} at the prime mover's RPM, or in this case:

$$WK^2_{EQ} = WK^2_{pm} + WK^2_{Red.} \left(\frac{\text{Red. RPM}}{P_M \text{ RPM}} \right)^2 + WK^2_{Load} \left(\frac{\text{Load RPM}}{\text{PM RPM}} \right)^2$$

Note: reducer RPM = Load RPM

$$WK^2_{EQ} = WK^2_{pm} + WK^2_{Red.} \left(\frac{1}{3} \right)^2 + WK^2_{Load} \left(\frac{1}{3} \right)^2$$

The WK^2 equivalent is equal to the WK^2 of the prime mover, plus the WK^2 of the load. This is equal to the WK^2 of the prime mover, plus the WK^2 of the reducer times $(1/3)^2$, plus the WK^2 of the load times $(1/3)^2$.

This relationship of the reducer to the driven load is expressed by the formula given earlier:

$$WK^2_{EQ} = WK^2_{part} \left(\frac{N_{part}}{N_{prime\ mover}} \right)^2$$

In other words, when a part is rotating at a speed (N) different from the prime mover, the WK^2_{EQ} is equal to the WK^2 of the part's speed ratio squared.

In the example, the result can be obtained as follows:

The WK^2 equivalent is equal to:

$$WK^2_{EQ} = 100 \text{ lb.ft.}^2 + 900 \text{ lb.ft.}^2 \left(\frac{1}{3} \right)^2 + 27{,}000 \text{ lb.ft.}^2 \left(\frac{1}{3} \right)^2$$

Finally:

$$WK^2_{EQ} = \text{lb.ft.}^2_{pm} + 100 \text{ lb.ft.}^2_{Red} + 3{,}000 \text{ lb.ft}^2_{Load}$$
$$WK^2_{EQ} = 3200 \text{ lb.ft.}^2$$

The total WK^2 equivalent is that WK^2 seen by the prime mover at its speed.

Electrical Formulas

To Find	Alternating Current	
	Single-Phase	Three-Phase
Amperes when horsepower is known	$\dfrac{\text{HP x 746}}{\text{E x Eff x pf}}$	$\dfrac{\text{HP x 746}}{\text{1.73 x E x Eff x pf}}$
Amperes when kilowatts are known	$\dfrac{\text{Kw x 1000}}{\text{E x pf}}$	$\dfrac{\text{Kw x 1000}}{\text{1.73 x E x pf}}$
Amperes when kva are known	$\dfrac{\text{Kva x 1000}}{\text{E}}$	$\dfrac{\text{Kva x 1000}}{\text{1.73 x E}}$
Kilowatts	$\dfrac{\text{I x E x pf}}{1000}$	$\dfrac{\text{1.73 x I x E x pf}}{1000}$
Kva	$\dfrac{\text{I x E}}{1000}$	$\dfrac{\text{1.73 x I x E}}{1000}$
Horsepower = (Output)	$\dfrac{\text{I x E x Eff x pf}}{746}$	$\dfrac{\text{1.73 x I x E x Eff x pff}}{746}$

I = Amperes; E = Volts; Eff = Efficiency; pf = Power Factor; KVA = Kilovolt-amperes; kW = Kilowatts

Locked Rotor Current (I$_L$) From Nameplate Data

Three Phase: $I_L = \dfrac{577 \text{ x HP x KVA/HP}}{E}$

See: KVA/HP Chart

Single Phase: $I_L = \dfrac{1000 \text{ x HP x KVA/HP}}{E}$

EXAMPLE:
Motor nameplate indicates 10 HP, 3 Phase, 460 Volts, Code F.

$$I_L = \frac{577 \times 10 \times (5.6 \text{ or } 6.29)}{460}$$

$I_L = 70.25$ or 78.9 Amperes (possible range)

Effect of Line Voltage on Locked Rotor Current (I_L) (Approx.)

$$I_L @ E_{LINE} = I_L @ E_{N/P} \times \frac{E_{LINE}}{E_{N/P}}$$

EXAMPLE:
Motor has a locked rotor current (inrush of 100 Amperes (I_L) at the rated nameplate voltage ($E_{N/P}$) of 230 volts.

What is I_L with 245 volts (E_{LINE}) applied to this motor?

$I_L @ 245$ V. $= 100 \times 254V/230V$

$I_L @ 245V. = 107$ Amperes

Basic Horsepower Calculations

Horsepower is work done per unit of time. One HP equals 33,000 ft-lb of work per minute. When work is done by a source of torque (T) to produce (M) rotations about an axis, the work done is:

radius x 2 πx rpm x lb. or 2 πTM

When rotation is at the rate N rpm, the HP delivered is:

$$HP = \frac{\text{radius} \times 2 \pi x \text{ rpm} \times \text{lb.}}{33,000} = \frac{TN}{5,250}$$

For vertical or hoisting motion:

$$HP = \frac{W \times S}{33{,}000 \times E}$$

Where:

- W = total weight in lbs. to be raised by motor
- S = hoisting speed in feet per minute
- E = overall mechanical efficiency of hoist and gearing. For purposes of estimating
- E = .65 for eff. of hoist and connected gear.

For fans and blowers:

$$HP = \frac{\text{Volume (cfm)} \times \text{Head (inches of water)}}{6356 \times \text{Mechanical Efficiency of Fan}}$$

Or

$$HP = \frac{\text{Volume (cfm)} \times \text{Pressure (lb. Per sq. ft.)}}{3300 \times \text{Mechanical Efficiency of Fan}}$$

Or

$$HP = \frac{\text{Volume (cfm)} \times \text{Pressure (lb. Per sq. in.)}}{229 \times \text{Mechanical Efficiency of Fan}}$$

For purpose of estimating, the eff. of a fan or blower may be assumed to be 0.65.

Note: Air Capacity (cfm) varies directly with fan speed. Developed Pressure varies with square of fan speed. Hp varies with cube of fan speed.

For pumps:

$$HP = \frac{\text{GPM} \times \text{Pressure in lb. Per sq. in.} \times \text{Specific Grav.}}{1713 \times \text{Mechanical Efficiency of Pump}}$$

Or

$$HP = \frac{\text{GPM x Total Dynamic Head in Feet x S.G.}}{3960 \text{ x Mechanical Efficiency of Pump}}$$

where Total Dynamic Head = Static Head + Friction Head

For estimating, pump efficiency may be assumed at 0.70.

Accelerating Torque

The equivalent inertia of an adjustable speed drive indicates the energy required to keep the system running. However, starting or accelerating the system requires extra energy.

The torque required to accelerate a body is equal to the WK^2 of the body, times the change in RPM, divided by 308 times the interval (in seconds) in which this acceleration takes place:

$$\text{ACCELERATING TORQUE} = \frac{WK^2N \text{ (in lb.ft.)}}{308t}$$

Where:

N	=	Change in RPM
W	=	Weight in Lbs.
K	=	Radius of gyration
t	=	Time of acceleration (secs.)
WK^2	=	Equivalent Inertia
308	=	Constant of proportionality

Or

$$T_{Acc} = \frac{WK^2N}{308t}$$

The constant (308) is derived by transferring linear motion to angular motion, and considering acceleration due to gravity. If, for example, we have simply a prime mover and a load with no speed adjustment:

Example 1

PRIME LOADER	LOAD
$WK^2 = 200$ lb.ft.^{2}	$WK^2 = 800$ lb.ft.^{2}

The WK^2_{EQ} is determined as before:

$$WK^2_{EQ} = WK^2_{pm} + WK^2_{Load}$$

$$WK^2_{EQ} = 200 + 800$$

$$WK^2_{EQ} = 1000 \text{ ft.lb.}^2$$

If we want to accelerate this load to 1800 RPM in 1 minute, enough information is available to find the amount of torque necessary to accelerate the load.

The formula states:

$$T_{Acc} = \frac{WK^2_{EQ}N}{308t} \quad \text{or} \quad \frac{1000 \times 1800}{308 \times 60} \quad \text{or} \quad \frac{1800000}{18480}$$

$$T_{Acc} = 97.4 \text{ lb.ft.}$$

In other words, 97.4 lb.ft. of torque must be applied to get this load turning at 1800 RPM, in 60 seconds.

Note that T_{Acc} is an average value of accelerating torque during the speed change under consideration. If a more accurate calculation is desired, the following example may be helpful.

Example 2

The time that it takes to accelerate an induction motor from one speed to another may be found from the following equation:

$$t = \frac{WR^2 \times \text{change in rpm}}{308 \times T}$$

Where:

   T   = Average value of accelerating torque during the speed change under consideration.

t = Time the motor takes to accelerate from the initial speed to the final speed.

WR_2 = Flywheel effect, or moment of inertia, for the driven machinery plus the motor rotor in lb.ft.^{2} (WR^2 of driven machinery must be referred to the motor shaft).

The application of the above formula will now be considered by means of an example. Figure A shows the speed-torque curves of a squirrel-cage induction motor and a blower which it drives. At any speed of the blower, the difference between the torque which the motor can deliver at its shaft and the torque required by the blower is the torque available for acceleration. Reference to Figure A shows that the accelerating torque may vary greatly with speed. When the speed-torque curves for the motor and blower intersect there is no torque available for acceleration. The motor then drives the blower at constant speed and just delivers the torque required by the load.

In order to find the total time required to accelerate the motor and blower, the area between the motor speed-torque curve and the blower speed-torque curve is divided into strips, the ends of which approximate straight lines. Each strip corresponds to a speed increment which takes place within a definite time interval. The solid horizontal lines in Figure A represent the boundaries of strips; the lengths of the broken lines the average accelerating torques for the selected speed intervals. In order to calculate the total acceleration time for the motor and the direct-coupled blower it is necessary to find the time required to accelerate the motor from the beginning of one speed interval to the beginning of the next interval and add up the incremental times for all intervals to arrive at the total acceleration time. If the WR^2 of the motor whose speed-torque curve is given in Figure A is 3.26 ft.lb.^{2} and the WR^2 of the blower referred to the motor shaft is 15 ft.lb.^{2}, the total WR^2 is:

$$15 + 3.26 = 18.26 \text{ ft.lb.}^2,$$

And the total time of acceleration is:

$$\frac{WR^2}{308}\left[\frac{rpm_1}{T_1} + \frac{rpm_2}{T_2} + \frac{rpm_3}{T_3} + \ldots\ldots + \frac{rpm_1}{T_9}\right]$$

Or

$$t = \frac{18.26}{308}\left[\frac{150}{46} + \frac{150}{48} + \frac{300}{47} + \frac{300}{43.8} + \frac{200}{39.8} + \frac{200}{36.4} + \frac{300}{32.8} + \frac{100}{29.6} + \frac{40}{11}\right]$$

$t = 2.75$ sec.

Figure A

Curves used to determine time required to accelerate induction motor and blower

Accelerating Torques

$T_1 = 46$ lb.ft.	$T_4 = 43.8$ lb.ft.	$T_7 = 32.8$ lb.ft.
$T_2 = 48$ lb.ft.	$T_5 = 39.8$ lb.ft.	$T_8 = 29.6$ lb.ft.
$T_3 = 47$ lb.ft.	$T_6 = 36.4$ lb.ft.	$T_9 = 11$ lb.ft.

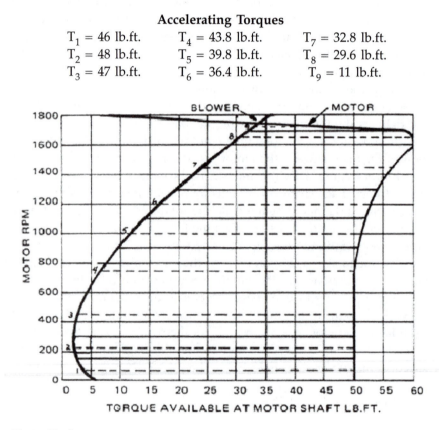

Duty Cycles

Sales orders are often entered with a note under special features such as:

"Suitable for 10 starts per hour"

Or

"Suitable for 3 reverses per minute"

Or

"Motor to be capable of accelerating 350 lb.ft.^{2}"

Or

"Suitable for 5 starts and stops per hour"

Orders with notes such as these can not be processed for two reasons.

1. The appropriate product group must first be consulted to see if a design is available that will perform the required duty cycle and, if not, to determine if the type of design required falls within our present product line.

2. None of the above notes contains enough information to make the necessary duty cycle calculation. In order for a duty cycle to be checked out, the duty cycle information must include the following:

    a. Inertia reflected to the motor shaft.

    b. Torque load on the motor during all portions of the duty cycle including starts, running time, stops or reversals.

    c. Accurate timing of each portion of the cycle.

    d. Information on how each step of the cycle is accomplished. For example, a stop can be by coasting, mechanical braking, DC dynamic braking or plugging. A reversal can be accomplished by plugging, or the motor may be stopped by some means then re-started in the opposite direction.

    e. When the motor is multi-speed, the cycle for each speed must be completely defined, including the method of changing from one speed to another.

    f. Any special mechanical problems, features or limitations.

Obtaining this information and checking with the product group before the order is entered can save much time, expense and correspondence.

Duty cycle refers to the detailed description of a work cycle that repeats in a specific time period. This cycle may include frequent starts, plugging stops, reversals or stalls. These characteristics are usually involved in batch-type processes and may include tumbling barrels, certain cranes, shovels and draglines, dampers, gate- or plow-positioning drives, drawbridges, freight and personnel elevators, press-type extractors, some feeders, presses of certain types, hoists, indexers, boring machines, cinder block machines, key seating, kneading, car-pulling, shakers (foundry or car), swaging and washing machines, and certain freight and passenger vehicles. The list is not all-inclusive. The drives for these loads must be capable of absorbing the heat generated during the duty cycles. Adequate thermal capacity would be required in slip couplings, clutches or motors to accelerate or plug-stop these drives or to withstand stalls. It is the product of the slip speed and the torque absorbed by the load per unit of time which generates heat in these drive components. All the events which occur during the duty cycle generate heat which the drive components must dissipate.

Because of the complexity of the duty cycle calculations and the extensive engineering data per specific motor design and rating required for the calculations, it is necessary for the sales engineer to refer to the product department for motor sizing with a duty cycle application.

Last Updated September 1, 1998

Copyright ©2007, Baldor Electric Company.
All Rights Reserved.

Index